Geopolitics in a Changing World

Geopolitics in a Changing World

Klaus Dodds
Royal Holloway, University of London

PRENTICE HALL

An imprint of PEARSON EDUCATION

Harlow, England · London · New York · Reading, Massachusetts · San Francisco · Toronto · Don Mills, Ontario · Sydney
Tokyo · Singapore · Hong Kong · Seoul · Taipei · Cape Town · Madrid · Mexico City · Amsterdam · Munich · Paris · Milan

Pearson Education Limited
Edinburgh Gate
Harlow
Essex CM20 2JE
England
and Associated Companies throughout the world

Visit us on the world wide web at:
http://www.awl-he.com

First published 2000

ISBN 0 582 27954 2

British Library Cataloguing-in-Publication Data
A catalogue record for this book is
available from the British Library

Library of Congress Cataloging-in-Publication Data
A catalog record for this book is
available from the Library of Congress

Typeset by 35 in 11/12pt Adobe Garamond
Produced by Addison Wesley Longman Singapore (Pte) Ltd.,
Printed in Singapore

Für Meine Grossmutter, Hildegard

Contents

List of figures

List of tables

Preface

A profound conviction that students taking courses in geopolitics and political geography have to possess a critical mass of historical knowledge and political awareness about not only recent events, such as the 1994 genocide in Rwanda and Burundi, but also the changing geographies of globalization, has been the motivation for my text. As a teacher of political geography, I believe that teaching materials should not only encourage further study but also invite a degree of self-reflection about one's own position in a seemingly abstract world of global politics. It is abundantly clear that some of the most exciting work in human geography sets out to make the connection between local acts of consumption, such as the purchase of food produce, and the wider circuits of capital, politics and knowledge. During the 1980s, the boycott of South African fruit and wine was considered to be a vital part of the isolation campaign against the South African apartheid regime. As Western consumers, we were urged not to buy Cape fruit and semillion from Paarl. Underlying that campaign was the notion that purchasing South African commodities (or supporting rebel cricket and rugby tours) was endorsing an implict acceptance of a particularly morally repugnant political regime. However, the lives and conditions of the predominantly Black South African workers affected by the Western consumer boycott were frequently marginalized in these struggles.

Various media forms such as advertising, books, films and television also participate in the construction and reproduction of particular political identities and representations of political space. Televisual and print advertisements for CFC-friendly aerosols freqently used images of the globe to stress that the purchase of these products was contributing to the environmental well-being of the earth. The shift away from CFC products was stimulated by the 1987 Montreal Protocol, which resolved that change was vital given the damage to the ozone layer in the earth's atmosphere. However, this shift away from CFC products in the North did not recognize that for many in the South a cheap substance used for refrigeration had been lost under this reduction process. While the Northern media and consumers may have believed that the Montreal Protocol was a global environmental agreement, it had profoundly unequal consequences for North and South. Using examples such as global environmental agreements, it will be demonstrated how space, politics, environment and knowledge interact within a range of spatial settings: local, state, regional and global. Moreover, it is imperative that the state is not considered in isolation from other organizations and institutions such as the international financial system, multinational corporations, non-governmental organizations and global forums.

The book uses a mixture of usual and unusual topics of political geography to examine how one might account for international political change and the implications for developing particular approaches to the subject. In recent years there has been a considerable blurring of boundaries between and within geopolitics, because it appeared senseless to consider economic, political and cultural realms of political life as conceptually and geographically separate from one another. The processes often associated with globalization have ensured, among other things, that academic interdependence is the new order of the day. Feminist political geographers, for instance, have demonstrated that gender relations (at a variety of geographical scales) have significant implications for evaluating international politics. Cultural geographers have considered how social and political life is constituted through discourses which manufacture meanings. In contrast to earlier studies, it is now widely accepted that meanings attributed to the political world are socially produced and contested rather than simply produced in a transparent and self-evident manner. Lastly, economic geographers have urged political geography to locate international politics within a context which recognizes the role commanded by changing economic conditions and non-state actors such as business firms and multinational corporations. More widely, disciplines such as international relations (IR) are contributing to a re-examination of key concepts in political geography, e.g. state sovereignty, territoriality and nationalism.

It is much easier to teach geopolitics now there are well-written textbooks which go beyond the standard topics of nations, territories and globalization to cover more unusual aspects such as gender issues, cultural survival and social movements; examples are P.J. Taylor (1993), Painter (1995), Ó Tuathail (1996), and Ó Tuathail, Dalby and Routledge (1998). Given my own interest in the academic and political intersection of geopolitics and disciplines such as international relations, I believe that some of the most interesting conceptual debates over state sovereignty and globalization have been developed and pursued in neighbouring research areas such as critical international relations theory and critical security studies. At the same time, I feel a degree of frustration that the contribution of human geographers to the investigation of global politics (such as the implicit territorial assumptions when studying international relations) has not been as significant within these related disciplines as one might have expected (Agnew 1994, 1998, Agnew and Corbridge 1995). Moreover, the investigation of popular sources concerning the varied representations of political space remains relatively neglected, considering the volume of material being generated by other subject areas such as cultural geography and feminist studies.

Structure of the book

The introduction examines some of the events and processes which appear to dominate the diplomatic agendas and television screens of the late 1990s.

Within this general context, Chapter 1 assesses the condition of geopolitics at the end of the twentieth century. Chapter 2 considers different theoretical approaches that might be taken to further explore the implications of globalization on geopolitics. Chapter 3 argues that the 'globalization of geopolitics' literature frequently neglects the profound inequalities between the industrialized democracies of the North and the developing countries of the South. Chapter 4 looks at the geopolitics of popular sources and argues that the debates surrounding geopolitics and globalization can also be considered through an examination of various media, institutions and even formal architecture. Chapters 5, 6 and 7 consider various events and themes that have been widely identified as globally significant: nuclear weapons and their proliferation (Chapter 5), global politics and the environment (Chapter 6), human rights and humanitarian intervention (Chapter 7). Chapter 8 is a short conclusion. Each chapter contains suggestions for further reading, and the glossary provides details on key terms such as globalization, geopolitics and the new world order. Within the text, glossary items are indicated in bold.

Klaus Dodds
London, December 1998

Acknowledgements

I owe a debt of thanks to Matthew Smith at Addison Wesley Longman, Paul Knox at Virginia Tech and Susan Smith at the University of Edinburgh for their constant encouragement, patience and constructive criticism. Over five years, students and colleagues at Royal Holloway, University of London, have contributed greatly to this book by providing me with information, ideas and feedback on the various courses that I have taught there. I am particularly indebted to Denis Cosgrove and David Simon for their specific comments. I am grateful to Justin Jacyno for dealing with my numerous cartographic requests. June Brain, Liz Young and Kathy Roberts at the Department of Geography, Royal Holloway, have also rescued me from countless near disasters on my computer. I would like to thank Rob Potter and Royal Holloway, University of London, for a sabbatical term in Michaelmas 1998, which enabled me to complete this book. Finally, thanks to all my friends from Bristol and London who have provided me with much humour and support. Any errors of fact and interpretation, however, remain my own responsibility.

Special thanks to Aloisius for casting an editorial eye over the final text. This book is dedicated to my mother and my brother, who have been an endless source of love and encouragement.

Klaus Dodds

We are grateful to the following for permission to reproduce copyright material:

The UN for figures 2.2, 7.1, 7.4, 7.5 and 7.6; Panos Pictures and Jon S. Paull for figure 3.1; Marcus Dodds for figure 4.4; *Reader's Digest* for figure 4.5; the Public Records Office for figure 4.7; Steve Bell for figure 4.8; Rex Features for figure 5.2; Panos Pictures and Heidi Bradner for figure 6.2; Panos Pictures and Howard J. Davies for figure 7.9; and Keo Ltd for the Keo beer label in the quiz.

Although every effort has been made to trace owners of copyright material, in a few cases this has proved impossible and we would like to take this opportunity to apologise to any copyright holders whose rights we may have unwittingly infringed.

Introduction

It has been widely suggested that the present century started and ended in Sarajevo (Thompson 1994). The murder of the Austrian Archduke Franz Ferdinand II by a young Bosnian Serb called Princip in Sarajevo, on 28 June 1914, precipitated the First World War. Subsequent events demonstrated the deadly consequences of violence on a massive scale. Eighty years later, Sarajevo again returned to the world's attention in 1992 on the eve of the violent disintegration of the federation of Yugoslavia. The destruction of multicultural and multi-ethnic Bosnia was seen by many European and American observers as a cruel indictment not only of a disappointing century but also, most disturbingly, of violent ethnic cleansing. In this context, Sarajevo is not only a geographical place but also a metaphor for a century characterized by industrialization, the modern state and war. In an alternative vein, the late Marxist historian E.P. Thompson referred to the twentieth century as an Age of Extremes, when humankind displayed an unprecedented capacity for destruction as epitomized by the Holocaust and the explosion of the nuclear bomb. Coincidentally, as Alex Danchev has noted, E.P. and his wife, Dorothy Thompson, spent a summer in Sarajevo in 1947 as part of a 20,000-strong socialist contingent intent on building a new railway between Sarajevo and Samac (cited in Danchev 1995).

In the summer of 1998, the Balkans again captured the attention of the world's media, as violence erupted in the former Yugoslav autonomous province of Kosovo; see Malcolm (1998) and Figure 0.1. It has been alleged that Serbian forces destroyed Albanian villages and towns in western Kosovo. The plight of the Albanian population has apparently worsened since the 1995 Dayton Accord, which brought about a ceasefire in Bosnia but left unresolved the status of this autonomous region. Kosovo, overwhelmingly populated by ethnic Albanians, has since the late 1980s been subjected to violent attempts by the Serbian forces to incorporate this region into the former Yugoslavia. It has been argued that the incidents in Kosovo are tragic examples of 'new wars', which overturn previous assumptions of struggles being based on land and stable ideologies, and ideas that foreign states refrain from interference in the 'internal affairs' of other states (Malcolm 1998). Increasing numbers of states and political leaders have recognized that the power of 'long-distance nationalists' (Benedict Anderson's phrase) can be considerable, as Greek and former Yugoslav communities in Australia, North America and Western Europe

Figure 0.1 Kosovo is populated by Albanian speakers (90 per cent of the total population) and the Kosovo Liberation Army has been fighting for independence from Serbia. Since 1997 some 250,000 Albanian Kosovans have had to flee their homes. In October 1998, European unarmed observers sought to maintain the peace between Serbian and Albanian Kosovan factions. An uneasy truce was agreed.

pressurize governments into taking lines of action. In contrast to previous conflicts, these 'local conflicts' are increasingly globally mediated by the demands of identity politics and ethnic cleansing rather than the Cold War struggle against communism or capitalism.

The difficulties with examining geopolitics in a changing world are inherent in the aforementioned observations about the twentieth century and Sarajevo. The notion of a twentieth century is a completely arbitary period of time based on Western and Christian values concerning timekeeping (Gould 1998). The birth of Christ and the invention of categories such as decades and centuries have played a significant role in shaping our (principally Western and Christian) common understandings of historical events. Within this temporal tradition, centuries have been extremely flexible resources. They have tended to be 'longer' than the customary hundred years (witness the description of the long sixteenth century from 1450 to 1640) and they have been 'shorter' (witness the description of the present century from 1914 to 1991). While most of the comments in this book are explicitly located within

Figure 0.2 The Dome of the Rock in Jerusalem was built in 691 and is one of the holiest sites in the Middle East. It is believed that Mohammed ascended to heaven from the rock at the centre, already revered as the spot where Abraham prepared to sacrifice his son Isaac. The dome is covered with 24-carat gold, marble and painted tiles. (*Photo*: Klaus Dodds)

a certain time frame, it is imperative to recognize that our categorization of time and space is culturally specific. Recent attempts, therefore, by British political leaders to excite public interest in the ongoing construction of the Millennium Dome in London have been treated with a widespread degree of scepticism and caution because it is by no means clear which temporal or spatial characterization should prevail within our analyses (Figure 0.2).

The fate of the former Yugoslavia, however tragic in the 1990s, is but one aspect of a changing world and represents only one illustration of the condition of humanity during either the twentieth century or the period following the ending of the Cold War (commonly taken as the period 1945 to 1990). Whereas most accounts of the new world order (1991 onwards) have been preoccupied with the regions of Europe, East Asia and North America, many

Figure 0.3 Complex humanitarian emergencies during the 1990s.

Anglo-American accounts of the new world order, for example, have had little time or space for 'Southern' regions, such as sub-Saharan Africa, unless they had a direct impact on 'Northern' foreign policies. For the African and Central American states crippled with debt burdens and environmental destruction wreaked by hurricanes (including Mitch in October and November 1998), the ending of the Cold War did not profoundly change the circumstances of their citizens. The failure of the post-Cold War United Nations to intervene during the height of the Rwandan massacres in 1994 (over 800,000 killed according to the Red Cross) confirmed the perceptions of many African observers that the American-dominated UN was not inclined to help Central African states which were widely perceived to be heavily indebted, ethnically unstable and strategically unimportant; see Prunier (1995) and Figure 0.3. This position contrasts starkly with the 1991 Persian Gulf crisis, where American willingness to gather together a multinational taskforce to liberate oil-rich Kuwait resulted in the mobilization of a massive military coalition. With the removal of the Cold War struggle between capitalism and communism, conflicts in Rwanda, Liberia and Somalia were frequently discussed in the context of the return of barbarism to a continent which had not (and possibly could not) integrated into the hegemonic system of states and global capitalism.

The existing preoccupations with the collapse of the Soviet Union and the rise of new forms of nationalism in Yugoslavia and elsewhere have highlighted both the fragility and persisting appeal of the nation state. It will be argued, however, that forces associated with globalization and regionalization seem to be stretching the traditional agendas of international politics. Territorial border and state power are not exhausted features on the global political agenda, and recent events in East Asia bear testimony to the significance of these issues. In the last few years, the People's Republic of China has threatened Taiwan with military action, shown little interest in negotiating over the cultural and political status of Tibet, reaffirmed its claim over the Spratley Islands in the South China Sea and secured the handover of Hong Kong in July 1997. The southern 'Russian' province of Chechnya proved to the world that former superpowers cannot handle men and women prepared to die for national self-determination. In December 1998, the beheading of four British workers held hostage in the Russian province further illustrated the failure of the Russian government in Moscow to protect human rights. The question of Israel and the West Bank remains unresolved, as yet another phase in the so-called peace process stumbles over whether the distribution of territory and administrative powers can be agreed upon between various negotiating parties.

The terms employed by academic writers and political pundits to actively reimagine the world without the presence of the Soviet Union and the Cold War drama deserve critical attention. Phrases such as 'new world order' and 'new tribalism' can frequently be either racist and/or ethnocentric. The frequent lament in the North by officials and diplomats for the loss of old certainties during the Cold War often effaces the fact that this was a struggle not only over different ideological conceptions of world order but also a material struggle for resource and territorial advantage in the developing world or **South**. The sheer bloodiness of the Cold War should never be forgotten in the midst of Northern academic and political rememberances for so-called old certainties. Everyday life could be nasty, brutish and short for those people living in areas of the world labelled 'developing' or identified as part of the 'Third World'. During the Cold War period, the Americans dropped 40,000 tons of bombs on Cambodia in the 1960s and 1970s; ignored the pleas of the people of East Timor to prevent a bloody invasion of their country by the Indonesian armed forces in 1975 after declaring independence from Portugal; undermined socialist governments in Chile and Nicaragua; illegally invaded countries such as the Dominican Republic (1965) and Panama (1989) and supplied weapons to countless vicious military regimes in Latin America, Asia and Africa (Pilger 1998). The record of the Soviet Union was not that much better either, in terms of its contribution to the overall welfare of people living in the developing world. The 'pacification' of places ranging from Afghanistan to the Yemen chillingly illustrates the levels of death and violence associated with the old certainties of the Cold War.

Phrases such as 'the return of tribalism' in the South to describe the state of the world after the Cold War are often emblematic of ethnocentric and/or racist geographical imaginations. The emergence of wars in Central Africa and the Balkans in the 1990s was often considered by Northern commentators to be symptomatic of a return to irrational tribalism. Bloody nationalism in Algeria, Bosnia and Liberia was underpinned by dangerous and unpredictable forms of ethnic hatred and religious fundamentalism. As Paul Smith noted:

> All problems would fall under the rubric of tribalism, zoned in the South or various pockets of barbarity, while the North would set about fostering and administering the increased integration of the millennial dream of a globalized, free-market economy. (P. Smith 1997: 9)

The North's interventions in places such as the Balkans and Haiti in the 1990s were based on the idea that fundamentalism and barbarism in the South (and/or the emerging capitalist world) could be safely contained in the so-called wild zones of the South. In contrast, terrorist outrages within Northern countries such as the 1995 Oklahoma City bombing and the 1996 Trans World Airlines (TWA) explosion over the North American Atlantic coastline were initially interpreted as the work of Islamic fundamentalists. Most mainstream commentators considered terrorist attacks in the wild zones of the North to be a product of individual evil and 'terrorist' states such as Libya, rather than as part of an ongoing series of material tensions and political struggles located in the global capitalist order. Ironically, the American justice system has just convicted a group of right-wing lunatics for the Oklahoma bombing and there is still some debate over the cause of the TWA airline crash (now believed to have been an electrical failure).

Negotiating geopolitics

Geopolitics should be attentive not only to the 'high politics' of states, intergovernmental organizations and territorial space, but also to a host of media, groups and other institutions which seek to represent particular visions or interpretations of political space. There are at least two good reasons for taking such a catholic approach to geopolitics. The first is concerned with traditional accounts of geopolitics, which have tended to focus on a remarkably limited range of materials and objects identified with international affairs. One of the most exciting and expanding developments in political geography has been the exponential increase in the type of material, places and issues embraced. If one reads the back issues of the journal *Political Geography*, it becomes abundantly clear that political geographers have used geopolitics with reference to a wide range of themes from popular literary sources such as *Reader's Digest* to the analysis of geopolitical motifs in films and other forms of media (Sharp 1993, 1996; Ó Tuathail 1996). The second reason why we should be sympathetic to an expanding conception of geopolitics is that the views and experiences of women, ethnic minorities and non-governmental organizations tended to be marginalized within traditional interpretations of states, environments and world politics. Recent work by feminist writers such as Cynthia Enloe have shown only too clearly that the world of geopolitics is not only gendered (i.e. at the very least the experiences of world politics affect men and women differently) but also heavily influenced by particular conceptions of the 'political', which have tended to emphasize the formal political spaces of the state and the international system (Enloe 1989, 1993).

The guiding principle of this work is that no one definition of geopolitics should be uncritically endorsed. All claims to define particular fields should be carefully examined in order to investigate, among other things, what is excluded and what is included. In doing so, it is suggested that traditional accounts of geopolitics have tended to be narrowly focused and excessively concerned with the geographical well-being of the nation state. In an age often characterized by globalization, it is apparent that the political world can no longer be defined primarily by the actions and activities of nation states. National cultures and politics are not containers that restrict and regulate flows of information, capital and people. International financial flows and transboundary flows of pollution, ideas and crime, plus the local activities of non-governmental organizations, merit the attention of geopolitical students because they emphasize the relational and mutually entangled 'messiness' of politics. So it is recognized that the identification of 'domestic' and 'international' politics is increasingly problematic in these changing times.

The ultimate purpose of this introduction to geopolitics is to contribute to the development of a critical approach to geopolitics (discussed in Chapter 2) together with a number of academic and political objectives:

- Geopolitics must acknowledge that the perspectives brought to bear on a particular subject area require close scrutiny.
- Geopolitics should recognize that investigating world politics requires a fundamental reconsideration of terms such as 'international' and 'political'. The expression 'world politics' is more inclusive because it highlights the concern for politics and political activities across the world rather than, as the term 'international' implies, an exclusive concern for interactions between nation states (Baylis and Smith 1997). This account is an exploration of how some areas such as environmental change and nuclear proliferation can illuminate the potential of critical analysis.
- The struggle to organize, occupy and administer territory has often been complex, contested and violent. Geopolitics must recognize that geographical information, whether in the form of electromagnetic maps, paper charts or virtual data, has frequently contributed to the struggle over space in global politics.
- Geopolitics should acknowledge that the capacity of people to influence the world around them depends on the availability of resources and knowledge. One apparent danger of focusing on the activities of the 'North' (as opposed to the 'South') or major powers such as the United States (as opposed to Chad) is that accounts of political geographical change are condemned for being partial and incomplete. The globalization of ideas and information is far from being global in terms of a capacity to create and access knowledge.
- Geopolitics should recognize that the development of the so-called 'borderless' world economy and the expansion of global media networks such as Cable News Network (CNN), have condemned some peoples and regions to apparent obscurity in the televised arenas of international politics.

East Timor: a forgotten injustice?

The people of East Timor have suffered persistent human rights violations since the illegal invasion by Indonesia in 1975 (Figure 0.4). With the absence of major television coverage, the massacre of up to 200,000 East Timorese failed to be subjected to widespread international condemnation. The one exception was the slaughter of 250 civilians by Indonesian forces at a cemetry

Figure 0.4 Since 1975 the people of East Timor have suffered persistent human rights violations at the hands of Indonesia.

in Dili, the capital; filmed by a British journalist in November 1991, it provoked international outrage when the footage was released. In spite of the Nobel peace prize going to the East Timorese peace campaigner Bishop Belo, the Indonesian government was not persuaded to withdraw from East Timor. Western powers such as the United Kingdom appear unwilling to compromise lucrative arms deals with the government of Indonesia. The 'visibility' of East Timor within global political space depended upon the activities of global media networks and their willingness to televise particular events and places. Bishop Aloisius Nobuo Soma, the retired Bishop of Nagoya (Japan), made the following observation in his address to delegates at the Asia–Pacific Conference on East Timor in 1994:

> Blessed are those who work for justice. The people of East Timor are working for justice, fighting for their rights, and they are blessed. . . . God is raising up people everywhere to walk alongside the East Timorese. . . . In 1989 Bishop Belo wrote that the world has forgotten East Timor. Let us show that is not true. (Cited in CIIR 1997: 38)

Further reading

This book and its themes are sympathetic to the other books within this series on human geography. For instance, the representational issues discussed in this book are also examined in *Ways of Looking at the World* and some of the economic issues are considered in *Economies in a Changing World.* At this point, I would recommend that any student interested in geopolitics should examine some of the current literature contained within journals such as *Political Geography* and *Society and Space.* There are also some excellent summaries on political geography and geopolitics, *The Geopolitics Reader* edited by G.Ó Tuathail, S. Dalby and P. Routledge (Routledge, 1998) and J. Agnew's *Geopolitics* (Routledge, 1998). Relevant material also appears in non-geography journals such as *International Affairs, Review of International Studies* and *Alternatives.* At the end of each chapter, there is a guide to further reading.

Chapter 1

Contemporary geopolitics

The Christian year 1998 should have been an occasion for global celebration in recognition of the fiftieth anniversary of the United Nations Declaration of Human Rights. In December 1948 Eleanor Roosevelt (widow of former US president Franklin D. Roosevelt) presented to the United Nations General Assembly the final version of a document prepared by the United Nations Commission on Human Rights. The subsequent declaration was adopted by the General Assembly (Chapter 7) and was widely considered as an important step towards a global consensus on the rights of citizens and the obligation of states towards its peoples. In the same year, the World Health Organization was established by the United Nations in order to promote better health standards for all the people in the world through extensive vaccination and disease prevention programmes. Three years later, the United Nations created the office of the high commissioner for refugees, based in Geneva, in an attempt to ensure that the dispossessed peoples of Central and Eastern Europe were protected by the international community (Figure 1.1).

Unfortunately, the year 1998 does not warrant uncritical celebration. There have been a few genuine achievements, such as the Easter Friday Agreement (April 1998) on the future of Northern Ireland and the subsequent award of the Nobel peace prize to the two prominent political leaders of the province, David Trimble and John Hume, in recognition of their efforts to end the conflict between Northern Irish Catholics and Protestants. It is hoped that this agreement will enable the two groups to enter a new phase of political, cultural and economic cooperation in close alliance with the governments of the United Kingdom and the Irish Republic. In many other parts of the world, human rights abuses and ethnic violence continue, ranging from illegal detention of dissident protesters to the widespread killing of peoples in Algeria, the Democratic Republic of Congo (formerly Zaire), Sri Lanka and the disputed Kashmir province in the Indian subcontinent. In these last years of the present century, solutions to the bitter wars, conflicts and disputes over territorial control, national identity and sovereignty remain as elusive as ever.

This chapter seeks to present themes that will permeate throughout this account rather than offer a comprehensive explanation of either recent civil wars or the major events that have shaped the last part of the present century. It will occasionally be Eurocentric or Atlantic-centric in the sense that the epicentre of globalization has frequently been located in the Euro-American

Figure 1.1 Delivering food aid to war-torn Sudan, a UNHCR relief truck. (*Photo*: Klaus Dodds)

region. However, the geographical focus of this book is such that a concern for global interconnection clearly entails consideration of other regions and spheres of interest, including the South, polar and oceanic spaces as well as areas labelled as a common heritage of mankind. Underlying this concern for globalization is a belief that understanding the historical and geopolitical contexts is an essential part of attempting to interprete some of the geopolitical changes (always likely to be provisional and contingent on future events) that have occurred in recent years.

The five major themes considered for discussion appear to be highly significant in the present decade and are best considered as tendencies and events, rather than clear-cut manifestations of processes associated with globalization, geopolitical change, technology and information exchange. In doing so, these themes touch upon dominant conceptions of time, the rise of financial and information networks, the fragmentation and resulting consolidation of the nation state and the rise of new geopolitical agendas in the aftermath of the Cold War.

- The end of the Cold War
- Information and financial networks
- Fragmentation and the 'sovereign' state
- Regionalism
- Media, humanitarian emergencies and war

The subsequent chapters engage these tendencies in a more detailed study of key political and geographical issues, such as environmental change, human rights and North–South relations.

The end of the Cold War

A number of commentators in the United States have attempted to explain the economic, geopolitical and cultural significance of the ending of the **Cold War**. The dicussions go to the heart of geopolitics because they are concerned with the political understanding of the world. The purpose of these reviews is to demonstrate that geopolitics should be understood as a project dedicated to the production, circulation and interpretation of global political space. Different forms of geopolitical interpretation are illustrated using the well-known commentaries of Francis Fukuyama and Samuel Huntington, concerning the passing of one geopolitical order and the emergence of a new (if uncertain) geopolitical order. While Fukuyama's thesis is decidedly more optimistic on the fate of the United States and the West in the new world order, Huntington's arguments on the interactions between civilizations are laced with apprehension and scepticism about the deterritorializing world order (Ó Tuathail and Luke 1994). However, the major theme unifying these two rather different commentaries is their profoundly *antigeographical* aspect, which tends to ignore the complex geographies of world politics. As with other influential American commentators, such as Robert Kaplan, diverse cultural regions of the world are frequently labelled as either barbaric or threatening to the interests of a great power such as the United States, thereby underestimating (perhaps deliberately) the inherent problems and issues in visualizing a rapidly changing world.

The most famous contribution – partly because of its timing, partly because of its title – was 'The End of History', an unashamedly triumphalist essay by Francis Fukuyama hailing the collapse of the **Berlin Wall** as a victory for the United States and a defeat for the forces of communism and tyranny. According to Fukuyama, the post-Cold War era would witness a transformation of regions such as Eastern Europe from state-managed communism towards liberal democracy and market economics. In his book, *The End of History and the Last Man*, Fukuyama returned to this theme of the triumph of liberal democratic political and economic systems:

> The most remarkable development of the last quarter of the twentieth century has been the revelation of enormous weaknesses at the core of the world's seemingly strong dictatorships, whether they be of the military-authoritarian Right, or the communist-totalitarian Left. From Latin America to Eastern Europe, from the Soviet Union to the Middle East and Asia, strong governments have been failing over the last two decades. And while they have not given way in all cases to stable liberal democracies, liberal democracy remains the only coherent political aspiration that spans different regions and cultures around the globe. (Fukuyama 1992: xiii)

Moreover, Fukuyama's thesis was predicated on the assumption that liberal democracy in alliance with market economies had the capacity to fulfil basic human needs of self-worth and material well-being. The consequences for the post-Cold War world would be

> the creation of a universal consumer culture based on liberal economic principles, for the Third World as well as the First and Second. The enormously productive and dynamic world being created by advanced technology and the rational organization of labor has a tremendous homogenizing power. . . . The attractive power of this world creates a very strong disposition for all human societies to participate in it, while success in this participation requires the adoption of the principles of economic liberalism. (Fukuyama 1992: 108)

Fukuyama's thesis on the global **hegemony** of liberal democracy and market economics has been heavily criticized by academic and political commentators who consider his arguments regarding the so-called Second and Third World as geographically and historically insensitive. While democratic governments have emerged in parts of Southern Africa, South America, and Southeast Asia, their long-term viability remains open to question given the repressive nature of administrations in Indonesia, Malaysia, Peru and South Korea. It is doubtful if these fledgling democratic states will follow the prescribed pathway or whether the previously authoritarian regimes succumb to the pressures of economic liberalism often produced by the demands for greater political freedoms, democractic elections, formal political participation and regular elections. With a population of over one billion people, the political and economic status of China must be considered to be 'out of kilter' with Fukuyama's prediction of the triumph of liberal democracy.

Fukuyama's arguments frequently underestimated the differences between liberal and democratic governments. The current administrations of countries such as Britain, Chile, Japan, the United States and India are characterized by different forms of institutions, participation and practices. The assertion that the West has been triumphant ignores the interplay of different cultures and civilizations which might lead to greater exchange and hybridity rather than global subservience to ideas of democracy and market economies.

By contrast, Huntington's paper 'The Clash of Civilisations', published in the American journal *Foreign Affairs*, was pessimistic rather than optimistic about the ending of the Cold War. Huntington argued that the new global order would be characterized by the interaction of seven or eight large civilizations: Sinic (Chinese), Hindu, Islamic, Japanese, Latin American, Orthodox, Western and possibly African (Figure 1.2). As the West is not considered to be culturally and politically dominant, the **new world order** will witness the growing influence of Islamic, East Asian and Chinese civilizations. The expanding Chinese economy has already contributed to the growing political confidence of the People's Republic of China. The handover of Hong Kong in July 1997 and the militant attitude (condemned by the Western powers) displayed towards the pro-independence forces in Taiwan in 1995–1996 are recent illustrations of China's political and cultural self-confidence. According to Huntington, the implication of such a development is that the West's capacity to manage world politics will be challenged by the Sinic (Chinese) and Japanese civilizations in the next century. In support of this proposition, Huntington presented evidence of a growing trend of anti-Western sentiment in Islamic and Asian countries, in spite of the fact that the Western-educated

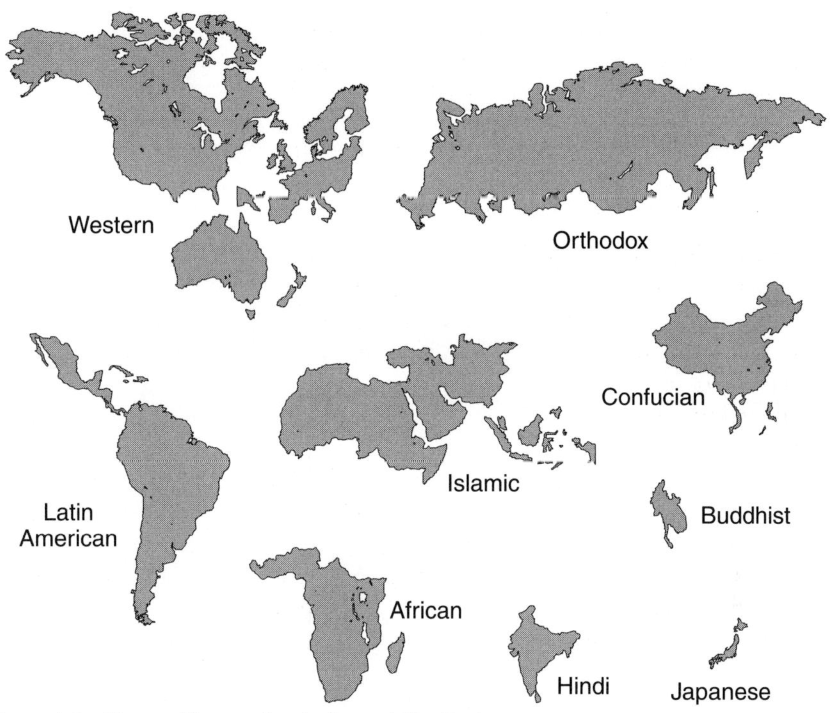

Figure 1.2 The world according to Samuel Huntington.

elites of these countries often speak English and are anxious to conduct business with international organizations.

Huntington's rejection of the Fukuyama thesis was further reiterated in his book *The Clash of Civilizations and the Remaking of the World Order*. As in his earlier paper, it is argued that the major divisions in the new global order will be based on the exchanges between civilizations rather than ideologies. For Huntingdon, the Cold War was a brief moment of international order in a longer historical context of confrontations and tensions. In the aftermath of the Cold War, international order will therefore be characterized by the return of civilizational tension as core states such as China, Japan and the United States seek to either expand or preserve their influence around the globe. His message for Western audiences is that the West will have to embark on a strategy which not only strengthens political and cultural values between Western civilizations but also seeks to construct new alliances with other civilizations (Figure 1.3).

As with Fukuyama's 'End of History' thesis, Huntington's arguments rely upon a series of sweeping generalizations about the state of world politics and processes such as globalization. The apparent threats posed by non-Western civilizations are frequently exaggerated and ignore the complexities posed by the apparent rise of fundamentalist movements in major religions such as Islam. Huntington's use of the recent disintegration of Bosnia and Russia (see Figure 1.4) to identify a tendency for multicivilizational societies to fracture

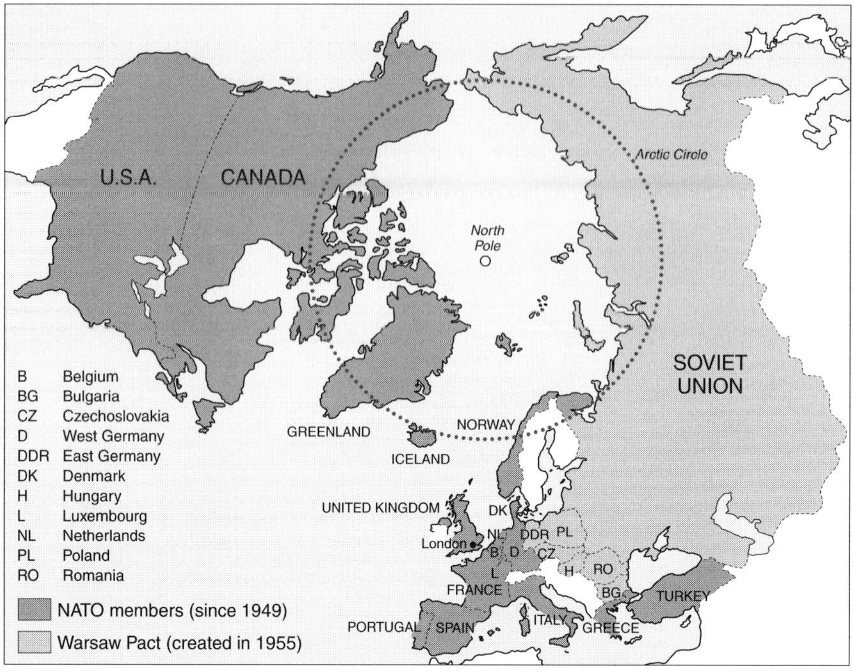

Figure 1.3 Euro-American security alliances during the Cold War.

Figure 1.4 The Kremlin, Moscow. During the Cold War, the political geography of the Soviet Union was frequently defined in the western media by single-image pictures of the Kremlin and Red Square. (*Photo*: Klaus Dodds)

along civilizational faultlines is a deeply flawed argument. While the example of post-apartheid South Africa provides evidence that multicivilizational societies are not inevitably doomed, the destruction of Bosnia between 1992 and 1995 cannot simply be reduced to the interplay of different civilizational forces (Islam, Orthodox Christianity and Western Christianity) because that would at the very least ignore Western complicity in the destruction of the former Yugoslavia. In a related vein, Huntington's characterization of the Latin American civilization as a backward, if reliable, appendage of the Western civilization effaces some of the important changes occurring within and between the Western and Latin American civilizations: large migratory movements of people and capital, hemispheric integration within the Americas and the growing inequalities between migrants and the settled population. The notion of a 'clash of civilizations' simply fails to capture the nature and intensity of interactions and exchanges between cultural regions. His all-encompassing definition of a civilization as 'the highest cultural grouping of people and the broadest level of cultural identity people have short of that which distinguishes humans from other species' has the effect of reducing people and events into broad and essential differences based on civilizational labels (cited in Ó Tuathail 1996: 244).

Both interpretations of the ending of the Cold War are geopolitical in the sense that they share a common concern for the mapping of global geopolitical space. Fukuyama, as a former member of the Department of State in Washington, advised on American foreign policy under the Reagan and Bush administrations; while Huntington's lengthy career spanned elite universities and political foundations such as the Council for Foreign Relations. Their commentaries formed a geopolitical discourse which construed international affairs as a dramatic stage for competing ideas and states.

Information and financial networks

The post-war international financial system based on the so-called Bretton Woods system of fixed exchange rates, currency convertibility and national controls on employment and interest rates has altered in the last twenty years. In the aftermath of massive surges in inflation, productivity decline and oil price rises, states began to inhabit a world composed of greater uncertainty *vis-à-vis* exchange controls, financial flows and declining powers of national regulation. In some cases, the emergence of new stores of currency (Eurodollars and petrodollars) led to the creation of novel geographies of international finance as stateless monies moved beyond the national regulatory powers of states and were located in offshore locations such as the Bahamas. In other areas, large economies such as the United States borrowed massively in order not only to survive the deep manufacturing recession of the 1980s but also to fund a substantial increase in the defence budget during the so-called Second Cold War (1979–1985). For countries in the developing world, their crippling debt burdens worsened due to the consumption by the United States of scarce international savings.

The net result of these dramatic transformations in the 1980s was to produce a greater opening up of the international economy and the financial markets. New communication technologies, such as computerized trading and electronic banking, have enabled leading centres of international finance, e.g. Tokyo, London and New York, to participate in an increasingly interconnected 24-hour global trading system. The globalization of finance and information networks in the late twentieth century has led many social scientists to question the capacity of nation states to control or even manage their international financial relations. In his book *The Vampire State and Other Stories*, Fred Block argues that the 'dictatorship of the international financial markets' and the activities of powerful private traders, such as the billionaire George Soros, are placing considerable pressures on national currencies and access to capital (Block 1996). In line with other observers, Block's adjective 'vampire' joins a long list of descriptions of the weakened nation state: hollow, phantom and defective. The significance of such terms lies in the way they question the effectiveness of the state in a world increasingly characterized by networks and flows rather than fixed territories and national spaces. As Manuel Castells has noted, information networks and the 'space of flows' are having a profound impact on the modern world economy (Castells 1996).

The growth of these flows and networks has undoubtedly been facilitated by media technologies and large media corporations, which have conspired to create a series of landscapes of interaction. Media industries led by moguls such as Rupert Murdoch, Silvio Berlusconi and Henry Luce have often been cited as examples of the growing influence of networks and access to communication. In 1989, for example, Luce and Warner Brothers merged in order to create a new company called Time-Warner, which had a staff of 340,000 and an annual turnover of US$18 billion. More recently, we have witnessed the current UK prime minister, Tony Blair, attempting to cultivate a close working relationship with media moguls because of their ability to generate news stories and influence public opinion. In the United Kingdom, Rupert Murdoch's News Corporation now controls 40 per cent of the newspaper market with key publications such as *The Times* and the *Sun*. Murdoch also controls the satellite broadcasting company Sky and has significant media interests in Australia, the United States and Asia. In the latter part of the twentieth century, it has been suggested that one of the main functions of the modern state is how to secure access to communication technologies and media networks rather than to protect territory. Indeed, the *Sun* newspaper reminded Tony Blair in May 1998 that its favourable coverage of the Labour Party during the 1997 general election campaign played a key role in their victory.

The pressures on the state in the 1990s are obvious when one considers how industrialized countries such as the United Kingdom and the United States struggle to ensure their currencies are not devalued by a range of international financial markets, media and speculators while simultaneously seeking to minimize their social spending programmes in order to improve accountability, efficiency and profitability. The geopolitics of money and currency devaluation tends to hit hardest on the poorest and most vulnerable states and communities, as powerful states seek to displace inflation, unemployment and excess production capacity onto weaker oppponents. The upsurge of leaner and meaner forms of state provision in the 1990s is undoubtedly related to the growing influence of international and transnational financial networks, which can place considerable pressures on state macroeconomic policies. Other observers have noted a wider trend towards a 'new public management revolution' in which new systems of governance have emerged for the monitoring of public expenditure, service efficiency and the role of the private sector in service provision (Desai and Imrie 1998).

In the South too, these financial networks are constraining state behaviour and national policy-making. In particular, it has been argued that the financial media, international bodies such as the **International Monetary Fund (IMF)** and **World Bank**, large-scale investors and multinational firms are contributing to an environment that effectively disciplines and monitors the behaviour of particular states. For states under the remit of the World Bank's **structural adjustment programmes** (SAPs), these constraints can be exacting as they seek to pursue good economic governance while avoiding substantial currency devaluations and interest rate rises.

The adoption of policies designed to liberalize trade and exchange rate policies left many Southern states vulnerable to transnational financial pressures when previous policies such as subsidies and price controls were deemed to be poor governance. This does not mean that these states are weak, defective and/or hollow, but it does explain the considerable capacity of various networks to influence government policies and strategies. While markets still rely on the long-term stability of governments and states, international financial flows can have a dramatic short-term impact on a country's economic and financial policies. The recent popular protests in Mexico, Venezuela and Indonesia against SAPs and currency devaluations have illustrated how the poorest people (often women and children) are frequently marginalized and victimized by international trends and government policies designed to rationalize the labour force and/or necessitate an increase in the price of basic commodities.

International organizations such as the **United Nations Education Scientific and Cultural Organization (UNESCO)** have campaigned for a reform in the ownership and management of media networks. In *Many Voices, One World*, UNESCO called for a more democratic ownership of the news media precisely because it recognized that the generation of news was massively unbalanced in the sense of being overwhelmingly produced by Northern industrial democracies (UNESCO 1980). Ironically, this call for a new world information and communication order has been greeted by an even greater centralization of media networks and news production in favour of a few American and English-language corporations such as CNN, Time-Warner and BBC World (Barker 1998).

Fragmentation and the 'sovereign' state

The processes of globalization and regional integration have to be considered against the important backdrop of geopolitical fragmentation. Tension between integrative and disintegrative forces could be said to be a defining feature of a century described as an 'age of extremes' (Hobsbawm 1997). It has been argued that globalization and fragmentation have ebbed and flowed throughout the present century (Clark 1997), and for much of this period the world was not only threatened by nuclear annihilation but also characterized by extreme levels of violence and a willingness to spend US$860 billion per year on arms procurement (1993 figure). Inequalities between North and South have widened in spite of the creation of unprecedented levels of wealth within the world community. Flagrant abuses of human rights have been accompanied by renewed endeavours to build upon the advancement of claims to human rights for all members of the international community. Above all, as Ian Clark states, 'the century was characterized by the greater interconnectedness of events on a global basis, while simultaneously being subjected to political processes of rupture and disintegration: it has been an age of globalization *and* [original emphasis] of fragmentation' (Clark 1997: 1).

During the 1990s, the economic dimensions of globalization coincided with a period of profound resurgence of ethnonationalism in Central and Southern Europe as well as Western and Central Africa. Globalization and fragmentation have impacted upon states and their policies in an often chronic manner. In 1991–1992, for example, four states collapsed and 22 new states were created. Two states (Germany and the Yemen) were also reunified. The destruction of Yugoslavia between 1992 and 1995 was assisted not only by a violent outpouring of nationalist violence and ethnic cleansing but also through the globalization of finance capital. The implementation of an IMF debt reform programme meant that the external debt burden of the federation was to be reduced through the slashing of state subsidies and public expenditure on health, education and welfare. Ethnic nationalism was in part a product of financial globalization. Food riots were accompanied by the creation of large reservoirs of unemployed labour as inefficient state-owned corporations were no longer protected by state subsidies. The harsh economic condition of Yugoslavia in the late 1980s later provided ample opportunity for ethnic leaders to recruit disaffected young men for their armies located in Bosnia, Croatia and Serbia.

Far from diminishing the power of nationalism, globalization appears to have heightened awareness of ethnic identities and insecurities about political life. Debates in the North about migration, immigration and trade have highlighted these sensitivities to globalization. Within the European Union (EU), immigration has emerged as one of the most controversial issues confronting states since the end of the Cold War. Fears that unrestrained migration would lead to the ethnic and cultural fragmentation of European states have provoked new legislation seeking to control movement from the Magreb (North Africa) and former Easter European communist states to the EU. Moreover,

Table 1.1 Asylum applications in the European Union, 1990–1995

Country	Applications	Granted refugee status
Germany	1,465,127	96,378
Sweden	205,986	114,502
Francc	199,222	59,499
UK	194,815	40,950

these immigration policies have been held responsible for the promotion of popular xenophobia in countries such as France, Germany and the United Kingdom. In the period between 1987 and 1992, there was evidence to suggest that a rising number of asylum seekers were in fact economic immigrants seeking entry into the EU (Table 1.1). In the main these groups of people were often employed within the informal sector of EU economies and the number of immigrants seeking asylum remains small (680,000 in 1992 for all EU nations).

In Germany, for example, far right groups have attacked hostels housing immigrant families from the former Yugoslavia and existing citizenship laws have prevented long-standing Turkish *Gastarbeiter* (guest workers) families from obtaining German citizenship. The rise of anti-immigration movements in France and Germany reflects a fear of cultural and ethnic fragmentation. The National Front Party of Jean-Marie Le Pen in France has frequently claimed that immigration from countries such as Algeria or Morocco is undermining the French Christian way of life. In 1995 it was estimated that 5 million Muslims were living in France out of a total population of around 60 million people. However, these figures tend to be taken out of context and used by centre right parties to argue that France has been overwhelmed by Islamic influences. Televisual and print media have also reported stories of the alleged widespread abuse of the asylum process and the welfare system in order to draw attention to the threat aimed at European cultures and identities. Immigrants and asylum seekers have become scapegoats for wider uncertainties facing the political and economic future of the European Union within a rapidly changing world. Yet, as many studies on immigration have suggested, the economic and cultural benefits of immigration far outweigh any additional demands that may be made on a state's welfare or health budgets. In 1990 a number of European Union countries such as France, Italy, Austria, Germany and Luxembourg signed the Schengen Agreement, which sought to tighten controls on immigration, visa control and asylum applications. The United Kingdom did not sign the agreement because it argued that its border controls were sufficiently robust. The EU also proposed that any future members of the EU, such as Poland or the Czech Republic, will have to pledge action to prevent illegal immigration as part of a condition for EU membership (Figure 1.5).

Within the European Union, changes in communication and transport networks are held to be responsible for undermining territorial and cultural

Figure 1.5 The European Union.

boundaries and creating new transnational communities and cultures. When the fatwa (a term used by some Muslims to describe someone or something that has offended the Islamic faith and therefore must be punished) against Salman Rushdie was announced in Iran, it transpired that the offending pages of the *Satanic Verses* had been faxed to the authorities in Tehran by a Muslim group from Bradford, West Yorkshire. For many ethnic groups, the ideal of the nation state retains a powerful attraction, as the struggles of the Kurds and the Chechens in the 1990s have demonstrated. It is also widely believed in the South that, in the face of further deterritorialization of the global political economy, statehood remains the best option for the purusit of economic and political security. Ironically, globalization may be actually strengthening rather than reducing the appeal of the nation state and separate territories.

Regionalism and the world economy

In the post-Cold War era, political scientists have sought to demonstrate how the political and economic integration of regions such as Western Europe had been based on the assumption that regional cooperation was a natural response to a world where there was no single sovereign authority. These regionalist forces have subsequently been identified by many political leaders as leading to the erosion of state sovereignty. In the face of rapid and unrestrained flows of finance and information, it was argued by some observers that states and societies were based on social networks rather than fixed entities and relationships located in time and space. The networks were considered to be open-ended and composed of permeable boundaries. Global flows of power, wealth and information placed new pressures on societies, states and communal identities, hence Northern states in particular resorted to promoting the rise of knowledge and skills as principal sources of competitive advantage by means of the new requirements of global capitalism and/or to defend state sovereignties through controlling the movement of capital and people. But according to Manuel Castells, the global system is increasingly frustrating these attempts to control and regulate spaces of flows (Castells 1996).

The realization that processes associated with globalization might promote the fragmentation of states and societies has led to a renewed interest in patterns of regionalism. For some observers, globalization was considered to be a desirable development because it could prevent the consolidation of hypercompetitive regionalism based on trading blocs in and around North America, Europe and East Asia (Gamble and Payne 1996). The emergence of rival trading blocs such as the European Union (EU) and the Asia–Pacific Economic Cooperation (APEC) was considered dangerous because it was argued that 'trading wars' could easily lead to 'hot wars' in the next century. Underlying this kind of argument was the assumption that regionalist tendencies in the world economy were no real substitute for the presence of a hegemonic power. It has been proposed under the so-called hegemonic stability thesis that a large economic and political power such as the United States could help to prevent countries from plunging into conflict by managing the current political world order (Hurrell 1998). By contrast, regionalism was often considered to be damaging because it promoted factional interests at the expense of the world political order.

For supporters of regional groupings, organizations such as MERCOSUR (Mercado Comun del Sur) help to consolidate South America's position within a highly unequal globalized world economy. The Treaty of Asunción in March 1991 initiated a new common market between Argentina, Brazil, Paraguay and Uruguay with the principal aims of promoting economic performance in the Southern Cone, improving communication networks and coordinating macroeconomic policies over issues such as tariffs. The Council of the Common Market and the Trade Commission of MERCOSUR have been charged with the consolidation of economic and political cooperation within the region. In 1995 a customs union was created at the same time as MERCOSUR nations expressed their support for the elimination of all trade barriers by 2006. Moreover, the hope was that MERCOSUR would be indicative of a growing confidence in the region and that democracy and market economies would have been firmly embedded. While there is no evidence that membership of MERCOSUR was considered limited to democratic governments, the association was swift to express support for President Wasmosy of Paraguay when he faced an armed uprising in 1994.

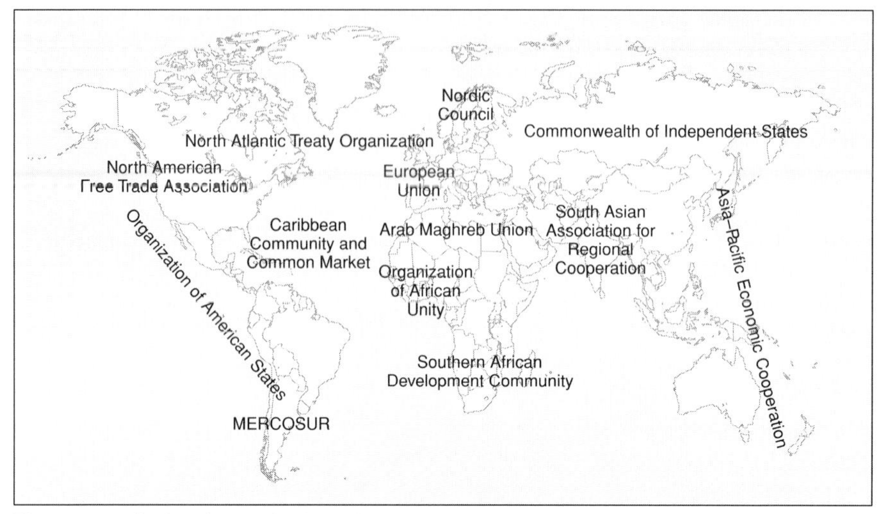

Figure 1.6 Regional organizations throughout the world.

Regionalism in the 1990s, in a variety of guises, is changing the nature of world politics. In the first place, regional organizations such as APEC and ASEAN have influenced Cold War geopolitical patterns based on superpower relations and/or divisions between a First World and a Third World (Figure 1.6). The Association of Southeast Asian Nations (ASEAN) was created in 1957 by the Cold War allies of the United States: Malaysia, Thailand, the Philippines, Indonesia and Singapore, and helped to promote economic and political development in the region. The ending of the Cold War witnessed old adversaries of the United States such as Vietnam and Laos being invited to join ASEAN in 1995. Ignoring Northern objections, ASEAN also admitted Myanmar (formerly Burma) to join in 1996, despite its poor human rights record. Meanwhile, the creation of a regional forum in 1994 enabled European Union states to engage ASEAN in trading and commercial discussions. Regional organizations are being situated within a global network of regional and subregional alliances, and North–South regionalism such as the APEC has contributed to a change in the international political landscape. Notwithstanding the arguments that the North American Free Trade Agreement (NAFTA) does not favour the political and economic interests of Mexico, this agreement is indicative of the new regional relationships which cross over boundaries between North and South. In East Asia, for instance, the roles of Japan and China in regional organizations such as the East Asia Economic Caucus changed and indeed replaced Cold War geopolitical and economic alliances.

The implications for international order and international society are difficult to ascertain, even though regionalism has contributed to more extensive notions of common values, mutual support and cooperation. It has been argued that there are five positive areas relevant to this discussion:

- Regionalism in the 1990s has encouraged local states to play a greater role in their regional security as international security in the post-Cold War era has been decentralized.
- Regional organizations can promote more effective agreements between states because of common interests, geographical proximity and local political pressure.
- Regional agreements have encouraged states to offer mutual support in the face of nationalist and ethnic conflict within a particular state or region. There would also be an added incentive to find local solutions because of common economic, political and cultural experiences.
- Regional organizations have been perceived as a mechanism for avoiding Southern marginalization in world politics as they can strengthen connections with the North and contribute to greater independence and self-sufficiency in the global political economy.
- Regional organizations can help to accommodate minorities and stateless peoples. The Scottish Nationalist Party remains a keen supporter of the EU because it believes that an independent Scotland could flourish inside Europe.

Nonetheless, sceptics of regionalism have pointed to the fact that regional organizations have a history of failing to live up to expectations. In the 1960s and 1970s, Latin American groups such as the Central American Common Market and Andean Pact failed to deliver effective macroeconomic cooperation. The failure of these regional groupings and others such as the Organization of African States appeared to validate realist views that the constraints posed by international anarchy and state cooperation are limited. Other regional groupings such as the Southern African Development Coordination Conference (1980) were based on a specific number of goals and managed to create a security consensus (based on reducing economic dependencies on South Africa and challenging apartheid) without the intervention of a superpower. In the midst of the Second Cold War, therefore, some forms of regionalism seemed more embedded than others.

The proliferation of regional organizations in the 1990s is undoubtedly posing new questions for world politics. With the ending of the Cold War, regionalism has emerged as a significant dimension in discussions of stability and order between regional forums and global or multilateral bodies. Within the United Nations, for instance, it has been argued that regionalism is complementary to the growing role of the United Nations in international affairs and that regional bodies could share the financial burden by and with greater involvement in crisis-ridden regions. UN involvement in West African states such as Liberia was helped by a regional coalition of peacekeeping troops led by Nigeria. In a similiar vein, the Cambodian peace process was assisted by Southeast Asian peacekeepers. Conversely, the attempts by the European Union to intervene in the former Yugoslavia were unsuccessful in terms of conflict resolution.

Media, humanitarian emergencies and war

The 1990s has been a shocking decade in terms of humanitarian disasters and genocide. During this period, television viewers have witnessed the ethnic cleansing of Kosovan Albanians and concentration camps in central Bosnia; large-scale massacres in Central Africa and rotting bodies floating in Lake Victoria; car bombings in Israel and Kenya; the destruction of villages in the disputed province of Kashmir; young children fighting in Liberia and Sri Lanka; starving people in Sudan and Mali; the systematic victimization of Kurdish people by Saddam Hussein and the devastating impact of Western smart bombs on Iraqi civilians (Weiss and Collins 1996). To geopolitical observers, it is increasingly apparent that the role and significance of television has given unprecedented opportunities to record the suffering and misery of distant and not so distant others (Chapter 4).

The growing importance of the televisual media in producing reports of events in distant places contributes to the way in which political leaders and societies construct visions of the political world (Shaw 1996). Television coverage of the fall of the Berlin Wall and the overthrow of governments in Eastern Europe in 1989 was said to have mobilized political support in Western Europe and North America for the democratic transformation of the region. It was argued that as a result of the intense television coverage of Eastern Europe, political leaders and business organziations were encouraged to redirect flows of aid and investment to this region at the expense of the South. Likewise, other commentators have pointed to the capacity of television coverage to produce superficial images of particular places. Public apathy in the United States and Western Europe towards the bombing of Iraq between January and March 1991 was in part created by television images of empty desert landscapes rather than a complex society composed of 80 million people. The lack of images depicting the fate and effect on the population gave the impression of war as a video game rather than as wholesale destruction. Whether a more detailed coverage of the Iraqi people and their suffering would have led more people to campaign against the bombing of the country must still be open to question.

The media portrayal of the 1991 Persian Gulf War (called Operation Desert Storm) has been studied in considerable detail ever since the first shots of tracer fire were shown streaming over the night-time Baghdad townscape (Kellner 1992). The quest for instant televisual news coverage by CNN during the campaign provoked some analysts to claim that this was a clear example of our globalized condition of demanding constant access to endless televisual images (Vattimo 1993). New forms of media and communications technology have transformed the speed and intensity of global political changes and, in doing so, they have challenged the ability of academics and political pundits to interpret and explain the condition of the world in the late twentieth century. The French social theorist Paul Virilio has proposed that the speed and intensity of globalization should be held responsible for the creation of new political conditions where the control of flows and borders is more

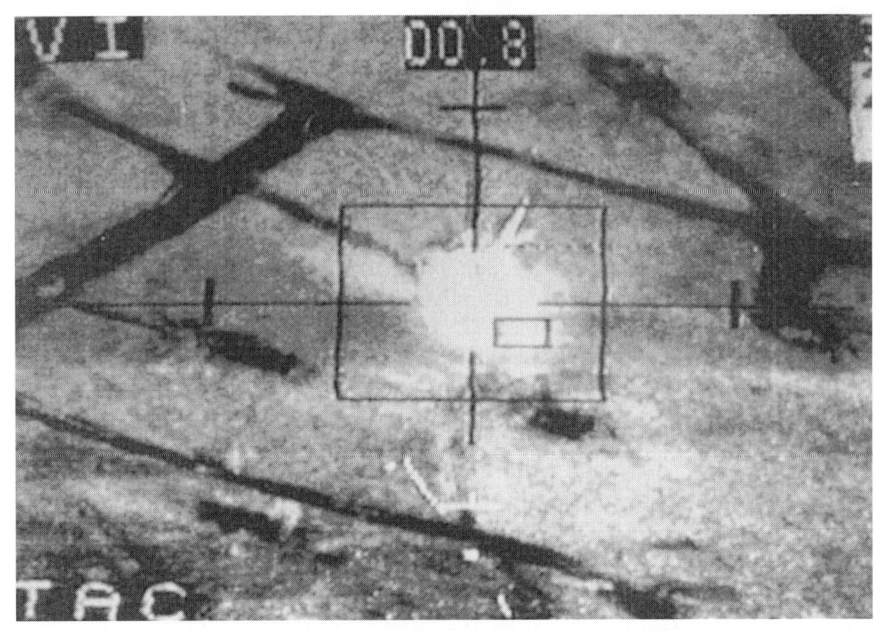

Figure 1.7 A video still taken from a camera inserted in a so-called 'Smart bomb'. The video images of the subsequent destruction became the defining image of Operation Desert Storm.

important than the traditional concern with sovereign places, impermeable borders and rigid geopolitics (Virilio 1986, 1989). The emergence of so-called chronopolitics (the control and distribution of time) has become more significant in world politics than the control and distribution of territory (geopolitics).

The creation of new cyberspaces has been cited as a reformulation of the problematic of geographical space. The power of the television screen to create and circulate visual spaces was the subject of an analysis of the 1991 Persian Gulf War (Der Derian 1992). The impact of simulation played a role in ensuring that Operation Desert Storm was played out not only on territorial space but also in virtual space, where geographical distance becomes irrelevant because reality is being produced on the screen. War becomes cyberwar and space is reduced to a series of computer simulations. Decision making during the campaign appeared to depend on data transmitted from satellites to the command centres in Saudi Arabia and the United States. The machinations of Operation Desert Storm were also reproduced in great detail on television screens (see Figure 1.7) within the sitting rooms of millions of Americans and other Western populations (Kellner 1995).

The distance between places around the world is increasingly mediated by the nature and speed of communication technologies. Television, according to some analysts, has led to the replacement of geographical space with simulated space. The consequence of such a mediation would appear to be a form of estrangement from not only a world composed of truth and certainty but also from others. It has been argued that the destruction of part of Iraq in January

1991, as seen through the medium of the television and video screen, actually made it more difficult for Western audiences to appreciate the suffering of their fellow human beings. The role and behaviour of Western field journalists did not assist the process of relating to distant others. As the British reporter Robert Fisk noted in his recollections of media reporting during Operation Desert Storm:

> Most of the journalists with the military wear uniforms. They rely upon the soldiers around them for protection. . . . They are dependent on the troops and their officers for communications and perhaps their lives. And there is a profound desire to fit in . . . [accompanied by] a frequent absence of the critical faculties. (Cited in Freedman and Karsh 1993: 424)

The capacity of television to generate forms of moral responsibility for the suffering of distant others has been a prominent feature of current debates on world politics. It is apparent that the response by the Western community to the televised suffering of others has been piecemeal and limited. Dramatic pictures from the former Yugoslavia also illustrated the difficulty for humanitarian workers and activists to cajole governments into particular courses of action in spite of harrowing pictures of concentration camps in places such as Omarska in Bosnia. Although the media made references to a latter-day holocaust, governments and international agencies struggled to coordinate a response to a conflict that had cost 130,000 lives and forced 2–3 million to flee their homes between 1991 and 1992 (Gow 1997).

Conclusion

Some of the earliest geopolitical writers, such as Halford Mackinder and Alfred Mahan, were concerned with the nature of interconnection between the earth and the challenges posed by the interstate system and international boundaries. In the twentieth century it has been recognized that the world's regions have become ever more interconnected: the expansion of the international system to encompass continents such as Africa and Asia; the growth of global markets; the incorporation of the polar regions and the ocean floors into mainstream international political agendas; the development of an international financial system in the aftermath of the Second World War. Perhaps the most striking illustration of the earth's oneness was provided by the publication of Apollo 17's photographs of the earth taken from space in December 1972. As Denis Cosgrove has noted: 'It (one particular image) captured, centreframe and with perfect resolution, the full terraqueous disk without a shadow or "terminator". The whole earth, geography's principal object of study, has been photographed by a human eyewitness' (Cosgrove 1994: 270).

The ways in which we represent the world can influence or reflect subsequent actions. There is no neutral or objective way of looking at the world, because our representations are always strategic and selective in the sense that some parts or objects of the world are emphasized more than others. Denis Cosgrove has argued that even terms such as 'earth', 'world' and 'globe' carry different cultural and political connotations (Cosgrove 1994). The term 'earth' is frequently used to describe a physical entity devoid of human beings, whereas 'world' has been used as an expression of order within a whole. The term 'globe' is usually intended to convey a sense of wholeness and is symbolized by reference to the earth. Each representation of the political and/or physical world has a particular perspective depending upon the author of that representation and his or her geographical location, economic circumstance, political affiliation, and so on. In these highly globalized times, representing the world along the lines of nation states and national interests becomes all the more difficult or complex when every practical feature of state sovereignty, such as the capacity to determine national laws and foreign policy, is open to challenge by other bodies of governance. The significance of these observations lies in the appreciation that each representation should not only be evaluated critically in the sense that there is no one correct or accurate view of our many worlds but also in terms of probable inequalities of social and political power. Therefore, any ensuing discussion of globalization places emphasis on the fragility of explanation due to the difficulty of explaining such a diverse and complex topic.

The processes of globalization, regionalism and geopolitical fragmentation are exposing new challenges for geopolitics. The hegemony of the nation state (in alliance with an international system) is being challenged by a range of developments, and the role and function of states as institutions and as forms of governance become changed. The growth of multilateral organizations, international agencies and multinational corporations has challenged the capacity of the state to formulate and implement legislation. The management of national economies has had to be carried out in a context where the wishes of state elites coexist with the demands of international money markets, international obligations and globalized flows of capital. Issues such as inflation, environmental damage, drugs and unemployment are transboundary in the sense that no one state or grouping of states can control these concerns. New forms of international cooperation have been required, such as the European Union's Social Chapter, which sought to regulate social and economic affairs within fifteen European states. Non-governmental organizations have challenged the functional capacities of states as globalization is thought to have increased the range of actions available to small groups and firms. For some analysts of globalization, the expanding influence of small groups in economic and environmental affairs may not be a positive step, because it could stimulate new forms of tribalism as opposed to globalizing shared human responsibilities (Naisbitt 1995; Simai 1997).

The significance of globalization for geopolitics and international relations should be apparent in terms of the implications for state sovereignty. Current debates over the so-called end of sovereignty have been most vigorously pursued in regions where the state has been strong (e.g. Western Europe) compared with some post-colonial states in Africa and Asia, which were never truly sovereign in the sense that they often lacked the capacities to control their territories internally. In the latter part of the twentieth century, European and

American observers have frequently noted that the means of states to control and influence events inside and outside their territorial boundaries has been challenged. According to some observers, the speed, intensity and acceleration of globalization questions modern conceptions of sovereignty, conceptions which depend on a clear demarcation of territorial space but which do not imply that state sovereignty has been eroded altogether. On the one hand, the modern conception of sovereignty makes it extremely difficult to contemplate alternative political communities other than those based on the state. Challenging conventional understandings of the state and sovereignty requires one to think not only of alternative forms of community and political organization but also how traces of state power coexist with forces that seek to undermine national sovereignty. On the other hand, state sovereignty remains influential precisely because it provides a convincing answer to the question of political identity. People and state elites are prepared to fight to the death for state sovereignty.

Further reading

For information on the United Nations and its various bodies, contact the Department of Public Information, United Nations, New York, NY 10017, USA. There is also a United Nations Information Centre at 21–22 Millbank Tower, London. For useful summaries of world politics in the late twentieth century, consult *The Globalization of World Politics* edited by J. Baylis and S. Smith (Oxford University Press, 1997), I. Clarke's *Globalization and Fragmentation* (Oxford University Press, 1997) and *An Unruly World?* edited by A. Herod, G. Ó Tuathail and S. Roberts (Routledge, 1998). On Western conceptions of time, see S.J. Gould's *Questioning the Millennium* (Vintage, 1998).

Chapter 2

The globalization of world politics

Globalization has become *de rigueur* in the 1980s and 1990s within popular and academic circles. In the business and management world, it signifies an apparently 'borderless world' in which trade, commerce and money can enjoy unimpeded movement over space and through time. For some political scientists, globalization conjures up a world in which the interstate system and state sovereignty have been challenged, perhaps even fatally undermined, by the transboundary flows of people, ideas and finance. In a more popular cultural vein, these processes and flows appear to signal the emergence of a McWorld, based on the widespread adoption and consumption of particular products such as McDonald's hamburgers, Nike shoes and Levi jeans. In short, globalization signals a massive sea change in the lives and conditions of citizens and states.

But there are also those who are cautious if not actually dismissive of the transformative powers associated with globalization. Some commentators argue that the phenomenon of globalization is simply exaggerated, not least because supporters tend to overemphasize the demise of the nation state and national boundaries (e.g. Hirst and Thompson 1996). From a Southern perspective, globalization often appears to refer to processes and flows that benefit rich, industrialized Northern democracies at the expense of regions such as Central America and sub-Saharan Africa (Chapter 3). The Internet, the mobile phone and Nike's France '98 advertising during the World Cup; this world probably seems incongrous to people simply struggling to survive on a daily basis. This has prompted a current of critical opinion arguing for globalization and its impacts to be considered alongside a view that recognizes profound differences and inequalities.

This chapter reviews the most significant approaches to world politics in terms of their response to globalization, beginning with some observations on theories. Initially, it considers the field of critical geopolitics; then comes a section on the body of political thought called political realism, highly popular within academic and policy-making circles, especially in the United States. Thereafter, liberal approaches are explored, because they seek to combine liberal and realist ideas on world politics. Finally, we return to some of the recent attempts to reconceptualize geopolitics, global politics and globalization. While this chapter does not consider in detail bodies of thought such as world systems theories and feminist approaches to politics, it nevertheless aims to demonstrate that a variety of approaches can be brought to bear on world politics. Subsequent chapters draw upon these other literatures in order to demonstrate that there are no right or wrong perspectives, because each depends upon specific conceptions of the 'political' and the 'geographical'.

Theories and world politics

In a world where there is much to know, there are also many ways of knowing. Claims to one particular way of knowing have frequently been exposed as either misrepresenting or excluding a variety of histories, places and contemporary experiences. Feminist commentators have been highly influential in exposing the fiction that there is a particular place from which one could get a detached overview of the world. This does not mean, however, that we are unable to make any kind of meaningful statement about the world around us. Rather it implies that we need to consider very carefully the intellectual and epistemological basis from which we make our claims about the world.

Explaining contemporary world politics is an extremely complex proposition not only because the range of materials available is susbtantial but also because the scope of interpretation is wide-ranging. Within the social sciences and humanities, it is now generally accepted that all forms of explanation are, in some sense, theoretically based. For the purpose of this chapter, it is assumed that this theoretical 'base' refers to an acceptance of a particular subject matter which enables a choice to be made between which issues or facts matter and which do not. Hence theory is not considered to be an option just because we are ignorant of the source from which a view of the world has been inherited. As Robert Cox once noted, 'theory is always for someone and for some purpose' (Cox 1981, cited in Gamble and Payne 1996: 6). The challenge for students of geopolitics and world politics is, among other things, to state as explicitly as possible these theoretical assumptions about the world.

It is only comparatively recently that formal academic disciplines such as international relations (IR) have been established in universities and institutions. International relations was created in 1919 within the Department of International Politics at the University of Wales, Aberystwyth. One of the founders of this discipline, David Davies, argued that international relations would help to prevent the future outbreak of wars because the scientific study of world politics would highlight the causes of political problems and would therefore contribute to the peaceful resolution of global tensions. In the immediate aftermath of the First World War, international relations scholars devoted much energy and time to investigating new forms of conflict mediation and the promotion of new international institutions such as the League of Nations. Labelled idealists by their opponents, supporters of this approach developed a powerful normative element – a concern to promote a model for the world rather than a commentary on the actual condition of world politics. In contrast, the earliest writers of geopolitics were more concerned with the interaction of states and territories than with attempts to improve the condition of the world. Although there have been honourable exceptions, such as the American geographer Isaiah Bowman and his idealist geopolitical text on the world after the First World War (Bowman 1921). While there were important differences between some of the earliest writers on international politics and geopolitics, there was nevertheless a common approach which dominated the present century, namely *realism*. The intellectual and political contribution of realism will be compared with other geopolitical approaches to world politics, such as liberal institutionalism and globalization.

The crux of this chapter is that geopolitics and international relations as academic fields are composed and shaped by the interplay of the real world and various fields of knowledge. Understanding the political world depends to a great extent on how we define that world in the first place. A point which will be reiterated is that there is no single overwhelming consensus concerning geopolitics. Some commentators concentrate on the geographical significance of the nation state and the international system, while others focus on the globalization of world politics. What is important, regardless of which perspective one adopts, is the recognition that these positions will have implications for understanding and explanation.

Critical geopolitics and world politics

The term 'geopolitics' was first used at the end of the nineteenth century by the Swedish political geographer Rudolph Kjellen. Since its formal inception, geopolitics has enjoyed a contested and controversial intellectual history. Kjellen's definition of geopolitics as 'the science which conceives of the state as a geographical organism or as a phenomenon in space' found favour in inter-war Germany (cited in Parker 1985: 55). In spite of being condemned as an 'intellectual poison', geopolitics has been a 'travelling theory' *par excellence* in the sense that it has entered a wide variety of disciplines and geographical regions. Over the last hundred years, many attempts have been made to chart the complex history of geopolitics, but few have managed to capture the historial and political complexities of the field. Throughout the twentieth century, academic work on geopolitics has often been conflicting, contradictory and confusing because of the variety of approaches brought to the historical examination of this intellectual field and contemporary analyses of world politics (Ó Tuathail and Dalby 1998).

In his recent review, John Agnew argues that geopolitics at the turn of the twentieth century was inspired by a particular way of viewing the world (Agnew 1998). The invention of the term 'geopolitics' coincided with a certain modernist belief that it was possible to view the world in its totality. The earliest texts of geopolitics reflected the belief that the European observer possessed the necessary intellectual and conceptual framework for visualizing the world as an external and independent 'object'. The earliest innovators of geopolitics in Europe and America tended to view geopolitics as a form of geographical reasoning which stressed the capacity of states to act within a changing global arena. Geopolitics was therefore a decidedly state-centric enterprise in the sense that the nation state was paramount. Moreover, the

Table 2.1 Key differences between traditional geopolitics and critical geopolitics

Traditional geopolitics	Critical geopolitics
National sovereignty	Globalization
Fixed territories	Symbolic boundaries
Statecraft	Networks/interdependence
Territorial enemies	Deterritorialized dangers
Geopolitical blocs	Virtual environments
Physical/earthly environments	
Cartography and maps	Geographic information systems (GIS)

Source: Adapted from Ó Tuathail (1998: 28).

physical environment was frequently conceptualized as a fixed stage on which political events occurred, rather than a dynamic and shifting problem which influenced the very conceptualization of world politics.

The major difference between traditional geopolitics and the more critical approaches is that the more critical approaches promote an opening up of political geography to methodological and conceptual re-evaluation (Table 2.1). Composed of various strands of social theory, critical geopolitics has sought to problematize the ways in which geographical discourses, practices and perspectives have measured, described and assessed the world. The inspiration for critical geopolitics lay in a belief that traditional political geography had failed to disrupt the widespread 'depoliticization' of human geography in the 1950s and 1960s. Boundary studies, for instance, were concerned more with the function and typology of frontiers rather than the provision of a critical evaluation of their significance within the international system. Boundaries are central to the discourse of sovereignty as they provide, among other things, the means for a physical and cultural separation of one sovereign state from another. More recently, the pioneering work of the French geographer Yves Lacoste played a valuable role in contributing to an agenda which focused on the role exerted by geographical knowledge in consolidating military power and state-centric politics.

Yves Lacoste was professor of geography at the University of Vincennes in France during the 1960s and 1970s. His unhappiness with the academic and political state of geography prompted Lacoste and his colleagues to create a new journal called *Herodote* in January 1976. In a famous study entitled *La géographie, ça est, d'abord, à faire la guerre!* (Geography is first and foremost for the waging of war), he argued that geographical knowledge had contributed to military power and state-centric politics. In his analysis of the American bombing of dykes in the Red River in North Vietnam, Lacoste demonstrated that the geographical information gathered on the region was being used by American forces to target and destroy its food-growing potential (Lacoste 1973, 1976). Political geographers were reminded that the often violent relationship between geographical knowledge and political power is central to political and moral concerns. Lacoste broke with the apolitical aspects of French political geography and clearly stated his belief that geographical work should be located within ongoing political struggles and concerns. Later Lacoste used *Herodote* to raise a series of issues such as decolonization, immigration, Islamic politics and nuclear missiles. Many of his ideas concerning the role of geographical knowledge in informing foreign policy and military politics have been drawn upon within Anglophone critical geopolitics.

Geopolitics is also no longer considered to be the study of statecraft and the great powers (the management of international affairs and the ideas that have influenced the practices of diplomacy); instead it is now perceived to delineate an intellectual terrain concerned with and influenced by the interaction of geography, knowledge, power and political and social institutions. Critical geopolitical writers have argued that geopolitics is a discourse concerned with the relationship between power-knowledge and social and political relations. The adoption of this position leads authors such as John Agnew, Gearóid Ó Tuathail and Simon Dalby to propose that understanding world politics has to be understood on a fundamentally *interpretative* basis rather than on a series of divine 'truths' such as the fundamental division of global politics between land and sea powers. For the critical geopolitican, therefore, the really important task is interpreting theories of world politics rather than repeating often ill-defined assumptions and understandings of politics and geography.

Challenging conventional categories of international or global politics is part and parcel of a critical evaluation of the role of geographical knowledge and its influence on social and political practices. The emergence of critical geopolitics and geopolitical economy in the 1980s is an indication of political geography arising from an empiricist past (i.e. the facts speak for themselves) towards a theoretically informed field of inquiry (Ó Tuathail 1996). In alliance with critical theories of international relations, critical geopolitics has sought to develop theories of world politics which acknowledge the ambiguity, contingency and the uncertainty of the world we live in. Allied with other developments within the social sciences and humanities, critical approaches to world politics tend to share the post-modern scepticism that the world can be rationally perceived and interpreted through particular techniques.

The starting point for critical geopolitics is to argue that conventional perspectives on geopolitics and international politics ignore the taken-for-granted assumptions that underpin those positions in the first place. Critical thinking poses questions about how current situations come to exist or how power works to sustain particular contexts. Critical geopolitical writers, in constrast to realist observers, argue that the assumption of a detached and objective researcher recording the observable realities of international politics is fallacious. Far from being objective, the research perspective of realism often contributes to the presentation of a view which appears to legitimize the power politics of states. In contrast, critical approaches to world politics would suggest that unless one challenges or questions contemporary structures and power relations, then academic approaches run the risk of merely legitimating

Table 2.2 Theories of world politics: key themes

Realism	Liberalism	Critical geopolitics
• National sovereignty • States • Military power • Anarchical world	• National sovereignty • States and non-state organizations • Limited international cooperation	• Interdependence • Globalization • Networks and nodal points • Representations of global space

existing practices. Critical geopolitical scholars now acknowledge that their approaches to world politics are self-consciously situated within a body of conceptual and methodological assumptions about the world. The theories on world politics are not detached from the world we seek to describe and explain, and in acknowledging this point, critical theorists may then contribute to the development of practical ideas about how progressive social and political change can be promoted (Table 2.2).

The analytical framework of critical geopolitics is derived from a mixture of sources, including discourse analysis, international political economy, feminist approaches and post-modern social theory. The greatest influence on the literature of critical geopolitics has been the Foucauldian insistence that one must explore the power-knowledge nexus in discourse. As Ó Tuathail and Agnew noted, 'Our foundational premise is the contention that geography is a social and historical discourse which is always intimately bound up with questions of politics and ideology . . . geography is a form of power-knowledge itself' (1992: 198). In contrast to earlier writings, geopolitics is now not considered to be a neutral technique or device for viewing the world; instead it is seen as a discourse (a linguistic device) which can be employed to represent the world in particular ways. Thus, the first and most noteworthy source of critical geopolitics was derived from an investigation of the discourses of geopolitics and international relations. Such a position implies that perceptions of the world are derived from a series of assumptions, rules and conventions that are brought to bear by those seeking to explain events and circumstances.

Edward Said's work on Western representations of the Middle East provides one of the clearest examples of how Foucault's insights on discourse and genealogies of power-knowledge are suited to research. Said's famous book *Orientalism* charts the creation and evolution of a series of imaginary geographies which construct the Middle East within Western geographical imaginations. Using British, French and American literary sources on the Middle East (Orient), he argues that the Middle East's complex place and cultural characteristics were reduced to a few defining features such as the 'threat' posed to the Euro-American culture by Arabs and the Muslim faith (Said 1978). The book's concern for the 'distribution of geopolitical awareness into aesthetic, scholarly, economic, sociological, historical and philosophical texts' (ibid.: 12) has been extremely thought-provoking.

Discourses are seen to influence the rules and conventions by which political behaviour is structured, regulated and judged. Critical geopolitics has argued that discourses play a prominent role in mobilizing certain simple geographical understandings about the world which assist in the justification of particular policy decisions. Political speeches, for instance, offer possibilities for the promotion of certain ideas to influential actors in world politics. The use of symbolism, metaphors and tropes can be crucial to the shaping of political understandings of specific instances. A brief example may be helpful now.

American representations of Libya

Ronald Reagan's decision to label Libya a 'terrorist state' in 1985–1986 was an important prelude to a growing reassessment by the United States of the new threat posed by Islamic fundamentalism. The bombing of the Libyan capital, Tripoli, in 1986 was the culmination of an American belief that Libya had funded terrorist activity in the Middle East and Europe. Rather than being simply a piece of rhetoric designed to provoke a political argument, the political outcome was to damage US–Soviet relations just as they appeared to be improving. Widespread public support in the United States for the Tripoli bombing may well have been assisted by sympathetic media coverage and a growing trend in Hollywood to produce films locating threats against America in the Arab world.

In December 1988, a Pan American airliner carrying several hundred people from London to New York exploded over southern Scotland. After months of investigation, the American and British police named two Libyan airline officials suspected of this act of terrorism. After ten years of oil sanctions (against Libya) following American pressure on the United Nations, the Libyan leadership has agreed to consider that the two Libyan suspects might face criminal proceedings in the Netherlands under Scottish law. Although the US and UK governments have condemned Libya for supporting terrorist activity, it has been largely ignored that the Americans carried out an illegal bombing raid on Tripoli in 1986, which also killed innocent civilians. Moreover, in September 1998 President Clinton authorized illegal missile attacks on alleged Islamic fundamentalist bases and factories in Sudan and Afghanistan in response to terrorist bombings on American embassies in Kenya and Tanzania, attacks which killed hundreds of mainly African citizens.

Critical geopolitics argues, therefore, that geopolitics should be conceptualized both as a form of discourse and a political practice. In their investigation of the Cold War, Agnew and Ó Tuathail acknowledged the geopolitical reasoning of American political figures such as Ronald Reagan:

> Political speeches and the like afford us a means of recovering the self-understandings of influential actors in world politics. They help us understand the social construction of worlds and the role of geographical knowledge in that social construction. (Ó Tuathail and Agnew 1992: 191)

Geopolitics reconceptualized as a discourse and a form of political practice has several implications:

- Geopolitics should be considered as a political activity carried out by a range of political actors, not limited to a small group of academic specialists.
- Geopolitical reasoning employed by American statesmen during the Cold War points to the fact that 'unremarkable assumptions about place and their particular identities' can be highly significant. The assumed threat of the Soviet Union overwhelming American or Western civilization often drew upon long-standing geographical depictions of the Eurasian landmass being populated by Asiatic hordes intent on conquering European and Slavic peoples.
- The current distribution of power within the international system means that some states such as the United States are in a better position than others to influence the production and circulation of political discourses, hence they possess the capacity to shape geopolitical understandings of the world.

Critical geopolitics and geopolitical economy

The distribution of power within the international system is a major consideration for the geopolitical economy. The two political geographers most closely associated with the geographical relations of economic and political domination and dependence are John Agnew and Stuart Corbridge (Agnew and Corbridge 1995). Critical geopolitics and geopolitical economy share a number of considerations:

- States are not the only influential force in international politics. The activities of multinational corporations, non-governmental organizations and firms are considered to be of importance. It is abundantly clear that states not only have to operate within a world economic system where flows of capital and technology transcend territorial boundaries, but also in a world influenced by the activities of business corporations who operate in more than one country or region. Multinational corporations often enjoy considerable independence from particular governments, even if they might be identified as an American or Japanese firm.
- The presumption that states pursue so-called national interests often underestimates the importance of sectional interests, which may be represented as national interests for political reasons.
- Analyses of international politics often neglect patterns of economic relations to the detriment of the international political structure. International relations are thus reduced to a concern for the interaction between states through diplomatic and political arenas, instead of the reciprocal action between the world economy and the power of the state.

Geopolitical discourse participates in the construction of geographical significance for places and regions which can be linked to wider material interests. The capacity of the United States, as the largest economic and military power, to represent the world economy in particular ways (e.g. as an open and unrestricted arena of free trade) matters due to its influence on international financial organizations such as the World Bank and the International Monetary Fund (Herod, Ó Tuathail and Roberts 1998). That does not imply, however, that individual states such as the United States or Japan can determine an increasingly integrated global economy. The strategies employed by great powers to maximize specific interests are frequently 'scripted' through particular representations about the world economy and international politics. Geopolitical economy is therefore concerned, among other things, with the interaction between great powers, the international political system and economic processes.

The importance of discourse and representational practices has been a hallmark of critical geopolitics. Many papers and books have explored how foreign policy professionals and academics have depicted and represented global political space. The formal geopolitical reasoning of these individuals can then be seen as contributing to particular visions or vistas of world politics. Said's concern for imaginative geographies and representational practices has also stimulated interest in rethinking the formal disciplinary history of Anglo-American geopolitics. Many writers from Mackinder onwards have argued that the Western world has been threatened by the countries and regions of the 'East'. Mackinder's earliest geopolitical paper, 'The geographical pivot of history', identified a 'heartland' within the Soviet Union and claimed that Western powers would be threatened by Eastern powers. He noted that: 'Were the Chinese, for instance, organized by the Japanese, to overthrow the Russian empire and conquer its territory, they might constitute the yellow peril to the world's freedom' (1904: 430). In a similar vein, American geopolitical writers in the 1940s and 1950s argued that the threat from the East in the form of the Soviet Union threatened to imperil the free world.

Realism and the Westphalian model

The international lawyer Richard Falk coined the term 'Westphalian model' of world politics in order to describe a world allegedly characterized by the territorial sovereignty of states, an anarchical international arena, legal and political equality between states, the inherent right of states to use force in order to settle disputes, and limited cooperation between states. The term 'Westphalian' is derived from the 1648 Treaty of Westphalia, commonly held to have ushered in the role of European states in shaping international politics. The institution of the nation state was henceforth considered to be the premier political organization in European and later world politics. More commonly, however, academics and policy-makers substitute 'Westphalian' by 'realism'.

Political realism (realism) is widely recognized as the most influential body of literature associated with international relations and geopolitics. This approach to world politics should not be confused with the philosophical realism of Roy Bhaskar or Andrew Sayer. The premise for the earliest realist writers in the 1920s and 1930s was that the so-called idealists had misunderstood the nature of world politics. For the realist, the world was unpleasant and populated by generally selfish human beings intent on self-gratification rather than collective improvement. The principal political force within world politics was the nation state and the major determinant of international relations was the balance of power between states (Table 2.2). Moreover, the interactions of sovereign states occurred in an international arena characterized by anarchy rather than peaceful cooperation, because self-interested states were not subject to the regulatory authority of a supernational body. According to some realist commentators, the League of Nations was doomed to failure because it did not recognize that states were intent on maximizing their national interests at the expense of international cooperation. In the final analysis, it was argued that states would often rely on military force in order to achieve their own ends. The recent events such as the Iraqi invasion of Kuwait in 1991 or the American bombings of Sudan and Afghanistan in 1998 serve to consolidate this conclusion.

In contrast to idealism, therefore, realism was concerned with the apparent realities of world politics. Although this supposed concern with the realities of global political life did not guarantee that this approach was either commonsensical or neutral. As with other approaches to world politics, realism makes a series of assumptions about political life, human nature, the international system and the interactions of nation states. As Rob Walker has demonstrated, realist views of political life embody a fundamental contrast between life inside and outside the state (R. Walker 1988, 1993). Within the state it is possible to live the 'good life' and to become part of a society characterized by citizenship, community and culture. Outside the state, the notion of an international community of people is effectively abandoned and replaced by an interpretation which judges international relations to be dominated by war, violence, uncertainty and selfishness. The conception of community was not, however, abandoned altogether, as realists argued that cooperation between states was possible in an international society based on rules, laws and customs which moderated the behaviour of states. The creation of the United Nations in 1945 was widely considered to be a step towards the peaceful regularization of international relations.

Realist views of world politics have nonetheless been hotly disputed by those who believe that the Westphalian model does not capture the complexities of modern political life. Frequent criticism is levelled at the approach favoured by realists who tend to be more concerned with the role and behaviour of states, to the detriment of other international bodies such as firms, non-governmental organizations and international organizations whose contributions are often significant in shaping distinctive political agendas. The monopoly of violence perpetuated by the nation state has been used by realists to argue that the pursuit of military security is the primary objective of states. In doing so, the definition of security has frequently neglected to address other factors such as environmental, cultural and economic forms of security. This concern for the military and political aspects of security has led to the underestimation of social and cultural factors in shaping world politics. Realists tend to view the international system as an anarchical arena inhabitated by states (regardless of their social and cultural backgrounds) which function in an undifferentiated manner. The implication of this position is that realists believe the nature of the international system is unaffected by cultural variation, for example.

Scholars such as Hedley Bull and Rob Walker have argued that the realist depiction of the international system as an anarchical arena depends upon the representation of domestic political life as characterized by order and relative peace. This depiction of the international system reveals more about the power of dichotomization rather than the way in which categories such as the national and international interact with one another to produce particular understandings of political life. According to Walker, the challenge for students is to present an account of the international which is different from, but not a negation of, the national. Finally, the assumption by realists that there is either an independent or commonsensical way of seeing the world is epistemologically naive. The categories brought to bear on the world by realist analysis have implications for the understanding of politics, international society and territorial states. For example, an Islamic understanding of world politics would elicit a different reaction to the problem of international politics than that of the West.

In response to these criticisms, realist writers such as Robert Gilpin and Kenneth Waltz have sought to bring a theoretical and conceptual rigour to political realism. Since the 1970s, neo-realism has been a highly significant body of thought in international relations, which has fought to remove some of the unscientific methods of realism and the anecdotal usage of historical and geographical case examples. The inspiration for so-called neo-realism evolved through the changing circumstances of the world economy and politics. It was becoming increasingly obvious that the state was only one institution (rather than the main actor) within international politics and that a concern for interstate relations needed to be located within a broader political framework, which included non-state organizations and transnational relations alike.

The publication of Kenneth Waltz's *Theory of International Politics* (1979) was probably the most significant contribution to the development of neo-realism within Anglo-American IR. Using Karl Popper's conception of scientific method, Waltz argued that realism needed to be reformulated as a positive theory intent to produce law-like propositions for the international system. His aim, as Chris Brown noted, was to produce a theory of the international system rather than to account for all aspects of world politics (Brown 1997: 46). In his dense analysis of the international system, Waltz proposed that two possibilities exist for the international system: a hierarchical or anarchical system. The hierarchical system is considered to be a system composed of different kinds of units organized under a clear line of authority, whereas the anarchical system is composed of units which are similiar to one another. Hence, according to Waltz, the international system is anarchical (and has been since medieval times) in the sense that states cooperate with one another as equals in the absence of any form of world government.

However, Waltz's conceptualization of the international system is also conditioned by the existence of great powers such as the United States. As with realism, great stress is laid on the significance of powerful states such as the United States in maintaining order within the global system. As the economic and political hegemonic power, the United States created a basic political and economic framework for the post-war world based on the United Nations, the Bretton Woods agreement and American military power. Neo-realism places considerable emphasis on the structure of power within the international system and the impact that it exerts on the prevailing political order. For Waltz, the existence of two superpowers in the 1970s was considered preferable to three or more powers because of the capacity to impose stability on the global political order. In the process, the superpower confrontation of the Cold War transformed the world into one large strategic arena.

Many neo-realists would be sceptical of those who support the 'globalization of world politics' thesis because, as they would argue, the state remains the principal actor in an anarchical international arena. In spite of changes to the nature of world politics, states in the post-Cold War era remain committed to the pursuit of national interests and remain cautious in terms of cooperating with other states. Recent events such as the US and European reluctance to intervene in the Yugoslav civil wars (1992–1995) would seem to bear out this observation. In spite of the increased domain of cooperation, neo-realists believe that states retain a rational mindset, motivated by self-interest and self-preservation. In contrast to Kuwait and the oilfields of the Middle East, Bosnia was not considered to be strategically important by the international community, even though there was a desire to end the suffering of civilians.

In turn, the critics of neo-realism have pointed out that these approaches to international politics tend to be inherently conservative in terms of theory construction and political aspiration. Waltz's theory of the international system simply accepts the existence of anarchy rather than seeking to analyse the ways in which the construction of anarchy facilitates particular interests (Brown 1997: 56). In Robert Cox's terms, neo-realism is a problem-solving set of theories rather than a series of critical theories which seek to change particular situations (Dalby 1991). For critical theorists, neo-realism is an impoverished approach to world politics because it does not concern itself with human emancipation or the search for alternatives to the present political condition.

The position of political geography within the corpus of realism and neo-realism is therefore difficult to locate, because few political geographers have explicitly acknowledged their theoretical assumptions about the international system or politics. Yet at the heart of realist thought lie a series of propositions about world politics which can be summarized as follows: 'International politics, like all politics, is a struggle for power. Whatever the ultimate aims of international politics, power is always the immediate aim' (Morgenthau 1948: 27). Within conventional geopolitics, many writers argue that political life is dominated by the interaction of states in particular geographical settings. No political geographical writer compares to the status of the American Hans Morgenthau and his best-selling book *Politics Among Nations*, which listed the six major principles of political realism (Morgenthau 1948). It has been argued that the implicit assumptions of traditional political geography have been inspired by realist thought: 'As it [realism] informs a rather large and influential literature on geopolitics and military affairs, for example, realism has often degenerated into little more than an apolitical apology for cynicism and physical force' (R. Walker 1993: 107). Peter Taylor has argued that traditional geopolitical thinking in the mode of Mackinder and Mahan was inspired by a tradition of power politics within international relations (P.J. Taylor 1993). Mackinder's model of competing land and sea powers was inspired by his commitment to promoting British imperial interests in the face of overseas competition from Germany and Russia. The development of railways was considered crucial to the balance of power between imperial nations because it would allow traditional land-based powers such as Russia to control vast land areas through speed of travel. The identification of the Eurasian landmass as a 'geographical pivot of history' pointed to the geopolitical significance of particular territories in the struggle for control over the earth's surface.

Traditional geopolitics has also been underwritten by many of the assumptions of political realism concerning the nature of the international arena and the significance of state sovereignty and national interests. In contrast to realist analysis of international politics, however, political geography and geopolitics have focused on the power of the land and the sea to shape international relations. Classical geopolitical writers such as Mackinder endowed the 'heartland' with the potential to influence world politics at the expense of the so-called rimlands and outer crescents. Fixed assumptions about the geographical significance of places littered the geopolitical discourses of European and American political geographers. Geographical divisions were considered timeless and thus immune to human alteration. As the American political scientist Ladis Kristof once argued:

> The modern geopolitician does not look at the world map in order to find out what nature compels us to do but what nature advises us to do, given our preferences. (Kristof 1960: 19)

The capacity of human observers to influence understandings of world politics is diminished when the meaning of place and region was considered static rather than capable of change. Geography was reduced to the role of simply providing a territorial stage on which the interactions of states unfolded. Recent work within political geography has suggested that this is a very restricted view of geography which ignores the how and why geographic spaces and places are made significant through the processes of discursive construction.

In conclusion, realism has been frequently condemned for being an incomplete intellectual and political project. While it could be advocated that an approach which stresses the significance of states, war and national interest is admirable, realist presumptions about the interstate system and national behaviour do not account for many features of world politics. If, for example, the national interest of states is the primary concern for political leaders, then why do Nordic countries such as Sweden give substantial amounts of their gross domestic product (GDP) to the cause of humanitarian and developmental projects in the Third World? This should not imply that the state and state sovereignty are exhausted either as a concept or as a political and legal power, even though many authors have suggested that national boundaries and identities are progressively blurred by transnational flows and processes. Writers sympathetic to liberal approaches to world politics counter that realism fails to explain how the international system constrains and influences state behaviour through a series of conventions, treaties and international organizations such as the United Nations.

The 'UN charter model' of world politics

Richard Falk also coined the term 'UN charter model' to describe a world in which states coexisted with other social and political actors, cooperation was not limited between states, rules and regulations were used to eliminate unacceptable features of world politics such as genocide and war, and the territorial boundaries of states were blurred by transnational and supranational relationships. These series of assumptions are the foundation of the approach to world politics called *liberal institutionalism*. This is an intellectual compromise between liberalism and realism because while it is recognized that states and national interests are important features of the international system, it is proposed that a variety of others also share global political spaces such as the United Nations, intergovernmental organizations and non-governmental organizations (NGOs).

Liberal institutionalists contend that the international arena is not entirely anarchical. Although they would agree with realists that the sovereign state is the major organization within the international system, they would not necessarily accept that there are no checks or balances on the behaviour of states. A series of conflict-mitigating factors and transnational institutions ensure that states do not behave in a selfish and violent manner. These include a variety of intergovernmental and transnational regimes such as the 1959 Antarctic Treaty, which ensured that the polar continent has remained a zone of peace and a place for international scientific cooperation (Figure 2.1). The success of the Antarctic Treaty System is undoubtedly based on the fact that forty-three states agreed to temper their own national ambitions for the sake of peaceful cooperation and the environmental protection of the region.

The most important **intergovernmental organization** which seeks to promote international cooperation and peaceful exchange is the United Nations (UN). Under the 1945 Treaty of San Francisco, the UN was created by the international community in the hope that the anguish of the Second World War could be replaced by peace, dialogue and universal solidarity. The purpose of the UN was spelt out in the UN charter (111 articles), which defined the common goals for the world community, such as the implementation of particular moral values and standards for international relations. Signatories to the UN charter had to commit themselves to the peaceful resolution of disputes; the sovereign equality of all other members; the principle of collective security and a range of other social, political and cultural concerns (Whitaker 1997). The UN sought to maintain order and codify certain forms of behaviour as either acceptable or unacceptable. The UN charter also established

Figure 2.1 Antarctica has been a unique zone of cooperation largely due to an intergovernmental organization called the Antarctic Treaty System. The small Argentine base in Paradise Bay is a reminder of the importance of scientific research in the region. (*Photos*: Klaus Dodds)

the following bodies: the General Assembly, the Economic and Social Council, the Trusteeship Council, the International Court of Justice and the Secretariat (Figure 2.2 and Table 2.3).

For critics of liberal institutionalism, the performance of the United Nations is indicative of the difficulties inherent in this body of thought. During the Cold War, the role of the United Nations was effectively neutralized by five great powers who were the permanent members of the Security Council (China, France, the United Kingdom, the United States and the Soviet

Figure 2.2 Kofi Annan, seventh secretary general of the United Nations. (UN/DPI photo by Milton Grant; reproduced with permission)

Table 2.3 Secretary generals of the United Nations

Years in office	Name	Country
1946–1953	Trygve Lie	Norway
1953–1961	Dag Hammarskjöld	Sweden
1961–1972	U. Thant	Burma (now Myanmar)
1972–1981	Kurt Waldheim	Austria
1981–1991	Javier Pérez de Cuéllar	Peru
1991–1997	Boutros Boutros-Ghali	Egypt
1997–	Kofi Annan	Ghana

Union). Armed with the power of veto, these states habitually paralysed the UN and its executive orders often on the basis that particular UN operations or directives would interfere with their own strategic or political goals. The alleged sovereign equality of UN member states was frequently exposed as 'hollow' during the Cold War, when these great powers either ignored UN resolutions or violated the sovereign rights of Third World states. Although the American invasion of the Dominican Republic in 1965 was declared illegal by a UN resolution, this did not deter the United States from pursuing its own strategic objectives in the Caribbean. In other cases, key strategic allies of the great powers, such as Israel, were allowed to marginalize significant resolutions, e.g. resolution 242, which called for a 'just and lasting peace' in the Middle East after the 1967 Arab–Israeli War. As part of this peace process, Israel was supposed to withdraw from the so-called West Bank territory of Jordan but resolutely refused to implement this part of the UN resolution.

The UN charter's opening preamble also invokes 'We the people', while at the same time denying membership of the UN to non-state organizations and stateless peoples such as the Kurdish people in the Middle East. Many scholars have argued that an understanding of world politics can only be achieved by recognizing that 'political life' is not dominated by nation states. While liberal institutionalists recognize that NGOs, intergovernmental organizations and multinational corporations have significant roles to play, they still tend to overemphasize the role and scope of the state in their accounts of world politics.

Geopolitics and globalization of world politics?

Globalization has emerged as a central point of theorization and debate within the humanities and social sciences. Over the last ten years, internationalization was replaced by globalization because it was considered to be more helpful in analysing cross-boundary interaction. Considerable debate concerning the geographical scope, historical relevance and technological intensity of cultural, economic and political forms of globalization ensued (Waters 1995; Scholte 1997). Against such a backdrop, it is unsurpising that contemporary thinking about human affairs has acknowledged how we are all participants in a world of global connections (R. Walker 1988). But the evidence for and against globalization is manifestly disputed by the wide-ranging debate over its origins and significance (Robertson 1992; Held 1995; Hirst and Thompson 1996). For the supporters of globalization, recent changes in the world system are so profound that global politics, economics and culture have been radically altered. For the sceptics, however, the features ascribed to globalization are either exaggerated or insufficiently located within a longer historical process of world capitalist development. The sceptics conclude that a more careful analysis would reveal that the present levels of integration, interdependence and involvement of national economies and polities are not unprecedented.

According to the sociological writer Roland Robertson, globalization can be understood as a process whereby social relations acquire relatively distanceless and borderless qualities because the world is becoming a single and highly integrated place (Robertson 1992). He argues that there has been an active process of social system building at the global level for at least the last century and a half. The development of international trade and political cooperation has facilitated this evolution. Over time the global system has become more complex and interdependent because of time–space compression and the development of global consciousness. Time–space compression has enabled the creation of more intense interdependencies with the result that sudden changes in one part of the world can have implications for others. Around the world, the rapid popularity of the Nike running shoe in North America and Europe led to an increased demand in production, which in turn had implications for the workers who produced them in Southeast Asia. Other examples can be drawn from the environmental sphere, where unregulated industrial development and practices can adversely affect areas in other regions. The uncontrolled burning of the Indonesian forest in September 1997 forced citizens in Singapore to wear protective masks in order to avoid breathing poisoned air, and it caused international air traffic to be diverted.

The development of a global consciousness is related to time–space compression. Robertson argues that global consciousness has been facilitiated by developments in media communications, which allow people to participate in global discourses on world peace, environmental protection and/or human rights. Since the 1960s these sociological processes relating to globalization have intensified around the world. As the French social theorist Paul Virilio recently noted:

> And yet critical space, and critical expanse, are now everywhere, due to the acceleration of communications tools that obliterate the Atlantic (Concorde), reduce France to a square one and a half hours across (Airbus) or gain time with the Train à Grande Vitesse, the various advertising slogans signalling perfectly the shrinking of geophysical space of which we are the beneficiaries but also, sometimes, the unwitting victims. (Virilio 1997: 9)

Places and peoples are being drawn together into the socio-political space of others. This transformation has eroded the principle of state sovereignty in the sense that states and societies are experiencing greater difficulties than ever before in controlling their own affairs within their national territories. For the supporters of globalization, world politics has been or is being fundamentally changed.

Arguments in favour of 'strong globalization'

Economic transformations

Economic transformations in the world economy have meant that national states are losing capacity to control their own national economies. Interest rate changes in one economic region swiftly impact upon other regional components of the world economy. Currencies and commodities appear to travel across borders with very little interference from financial institutions and/or states. Within this apparently borderless world, business gurus such as Kenneth Ohmae have argued that the nation state is an outmoded institution which is ill-equipped to deal with world markets and borderless transnational corporations (Ohmae 1990). The currency crises in Russia, Brazil and Southeast Asia seem to confirm this observation, as states struggled to bolster their collapsing currencies in the midst of recession during 1998.

Information technology

Information and communication technologies have promoted the growth of global electronic networks, which enable information to be sent rapidly across the world. The development of the Internet in the 1980s is probably the most significant illustration of the global network society (Castells 1996). From a political perspective, nation states can often find it hard to control the flow of sensitive information. The British government's attempt to prevent the British publication of the book *Spycatcher* in 1986 was ultimately doomed because these memoirs of a former MI5 spy, Peter Wright, had already been published in the United States and Australia. At the time, television pictures in the United Kingdom showed how enterprising British citizens had travelled to the United States and returned to sell copies of *SpyCatcher* at street corners and in the forecourts of motorway (highway) service stations. The same is true of Kitty Kelly's book *The Royals*. It came out soon after the death of Diana, Princess of Wales, in August 1997 and although freely available in the United States and on the Internet, it could not be bought in the United Kingdom. UK libel laws were effective in thwarting UK publication of *The Royals*, but they could not prevent its publication elsewhere in the world.

Global risk

We live in a global risk society which has to confront transboundary dangers, such as AIDS, and environmental issues, such as pollution, often beyond the control of a single state or a group of states. Written during the aftermath of the 1986 disaster at the Chernobyl nuclear power station, Ulrich Beck's book *Risk Society* was a powerful account of how modern societies experience rapid and accelerating change in a host of fields, including information technologies and financial markets (Beck 1992). Beck's description of these changes identified 'risk' as central to our late-modern culture because so much of our thinking is 'what if' thinking in the face of uncertain futures. Unsurprisingly, relatively affluent people in the United States and Europe are now spending more money than ever before on insurance policies.

Regional organizations

The expanding influence of regional organizations such as the European Union are a reflection of a growing belief that neighbouring states have to cooperate with one another in order to secure the best possible position within the global political economy. Decision-making powers of the nation state are being superseded (in some cases) by regional bodies such as the European Council of Ministers. The UK government throughout the 1980s and 1990s was instructed by the European Council of Ministers to carry out certain initiatives (such as the culling of BSE-infected cattle herds in the mid-1990s) even though the UK parliament had voted against their full implementation.

Multinational corporations

The rise of multinational corporations means that new forms of global politics are challenging old forms of international politics. In short, states are having to compete with a variety of non-state actors, and this has often been perceived as detrimental to sovereign state power. Armed with sympathetic television coverage, non-governmental organizations such as Greenpeace have been highly effective in challenging the decision-making powers of governments. One clear example of this capacity to contest governmental strategies was evident in 1989 when Greenpeace, in alliance with other environmental NGOs, launched a global campaign against proposals to devise a minerals agreement for the Antarctic. Television campaigns against the minerals proposals were combined with mass public demonstrations in Australia, the United Kingdom, New Zealand, France, the United States and Canada. Within two years these proposals had been replaced with a new environmental framework which stressed that mining would be banned in the Antarctic region (Stokke and Vidas 1996).

Arguments against 'strong globalization'

Sceptics have lately argued that the features associated with globalization have either been overexaggerated or distorted. In their powerful critique of economic globalization, Hirst and Thompson (1996) contend that the present state of the world economy has not made states powerless in the face of rapid and uncontrollable flows of capital:

> The notion of globalization is just plainly wrong. The idea of a new, highly-internationalised, virtually uncontrollable global economy based on world market forces . . . is wide of the mark. (Hirst and Thompson 1996: 47–48)

They argue that the contemporary situation may not be as unique as many critics have suggested. Five major claims underline their critique of globalization:

- Economic activity continues to be nationally based in spite of the existence of the world economy and transnational flows of capital and commerce. In contrast to the suggestion that transnational corporations are dominating world trade patterns, it has been shown that they retain on average two-thirds of their assets in the home base and remain embedded in a particular national context. Within the major Northern economies, international business remains closely tied to home territory in the sense of overall business activity, location of sales, declared profits, and research and development.
- Globalization is no more than the sum total of international flows of trade and capital between countries, not an economic system that articulates on a global scale.
- Flows of trade and capital remain overwhelmingly concentrated in self-contained regional groupings such as the European Union, North America and East Asia (Figure 2.3). Hence contemporary flows of trade and commerce are not overwhelmingly global. The geographical reach of world capitalism has actually receded in terms of foreign capital flows and world trade, largely at the expense of sub-Saharan Africa and Latin America.

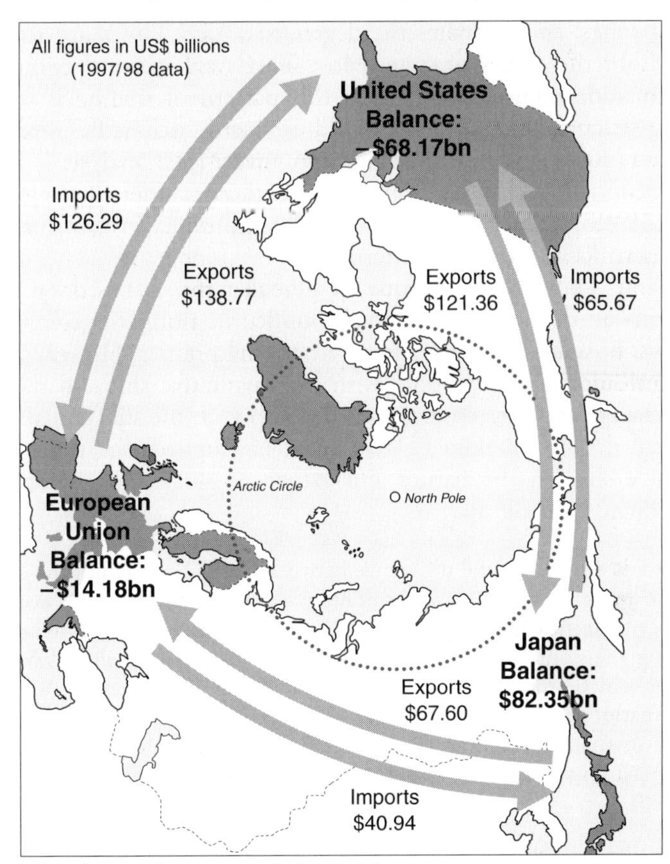

Figure 2.3 The three major centres of the world economy.

- National economic regulation is still possible, because there is scope for specialization and regulation through institutions and agreements. Pressure from above and below the nation state in the form of regional economic blocs (e.g. the European Union) alert us to the fact that national regulation has to coexist with other flows and forces.
- Most transnational companies (TNCs) are not global in the sense that their headquarters and major trading activities are concentrated in the three major trading regions (Europe, North America and East Asia) of the world economy. According to the index of industrialization, only twenty-one out of the top one hundred TNCs have a high globalized profile.

In support of these assertions, Hirst and Thompson suggest that the current trends in the global economy are actually similiar to the period between 1870 and 1914. The arguments pertaining to the uniqueness of the present world economy are thus considered to be overstated. States have had to deal with processes such as internationalization for a considerable time. Moreover, the prospect of transnational corporations taking over significant decision-making powers of states, overestimates the degree to which TNCs are actually independent from the affairs of state. Most TNCs are national companies who happen to trade internationally, and this pattern of trading is still concentrated in particular regions of the world economy such as East Asia, Western Europe and North America. Globalization, under their analysis, is found to be a phenomenon that is overwhelmingly Northern, rather than global, in the sense of the geographical distribution of foreign direct investment, communication networks and trading patterns.

Hirst and Thompson's critique of globalization is based on a series of observations on the world economy, grounded in a quantitative evaluation of trade flows, business activities and international politics. However, this thesis fails to deal adequately not only with the qualitative shifts in the nature of global exchange and interconnections but also with the substantial constraints on national decision making. David Held has argued that there has been a considerable shift in the nature and extent of global interaction compared with the nineteenth century:

> For there is a fundamental difference between, on the one hand, the development of particular trade routes, or select military and naval operations or even the global reach of nineteenth century empires, and, on the other hand, an international order involving the conjuncture of: dense networks of regional and global economic relations which stretch beyond the control of any single state . . . extensive webs of transnational relations and instantaneous electronic communications . . . a vast array of international regimes and organizations which can limit the scope for actions of the most powerful states; and the devleopment of a global military order. (Held 1995: 20)

The consequence of such a shift in the global order is that the nation states and national economies have to coexist with a range of networks and social actors. International politics has become more complicated in the late twentieth century as multinational corporations (accounting for two-thirds of the world's trade) contribute to the development and intensification of global circuits of production and exchange.

As a counter to these arguments concerning either the triumph of global capitalism and/or the calling into question of globalization itself, I would argue that globalization is probably best considered as an intensification of interaction between national and transnational social formations operating through the interstate system. Due to external influences, this means the state has lost some capacity to regulate a national economy through deregulation, exchange and interest rates, and fiscal policy. Increasingly, Northern states such as the United Kingdom and Germany have developed economic and political policies which promote global competitiveness, encourage inward investment and develop macroeconomic policies but remain susceptible to externally influenced interest rates and currency fluctuations. From a cultural perspective, it is abundantly clear that the world has not been reduced to a homogeneous cultural mass, despite the effects of vast flows of people, business and tourism. Traditional boundaries between territorial and social spaces have become blurred, and hybrid cultures and identities have been the defining feature of this interweaving of the local and the distant. As Anthony Giddens has argued:

> Globalization is not just an 'out there' phenomenon. It refers not only to the emergence of large-scale world systems, but to transformations in the very texture of everyday life. It is an 'in here' phenomenon, affecting even intimacies of personal identity. . . . Globalization invades local contexts of action but does not destroy them; on the contrary, new forms of local cultural identity and self-expression, are causally bound up with globalizing processes. (Giddens 1996: 367–68)

Seen from a political perspective, various writers have sought to convey a sense that transnational governance challenges state sovereignty over domestic affairs and the international system based on interstate diplomacy. In contrast to Hirst and Thompson, it has been argued that new forms of governance have emerged on the world stage, including governments and firms negotiating among themselves, transnational structures such as the United Nations, and the growing influence of non-governmental organizations within political spheres (see below). Other commentators have pointed to the growth of a global civil society where international social movements and the mass media contribute to a new civic awareness of human tragedies, environmental disasters, pollution, war and structural inequalities (Shaw 1996). The end result of these kinds of forum is not the irradication of the state and its power to regulate a national economy, but a reworking of national economic and political life in the context of transnational flows of capital, commerce and governance.

Conclusion

This chapter has concentrated on the idea that there is no single intellectual pathway for the comprehensive study of world politics. Although it has been critical of realist and neo-realist accounts of international politics, remember that the pursuit of national security is a major issue for some states and regions such as Palestine, Israel, Lebanon and the wider Middle Eastern region. Although recent interest in the so-called globalization of world politics has drawn attention to these interrelationships, critics of globalization have warned that some writers have overemphasized the declining power of the state and underestimated the fundamental differences that exist between North and South. However, other theorists argue that realism does not really consider transnational relations between states and non-state actors. As a result, it presents a rather restricted view of world politics, which fails to acknowledge the diffusion of networks and actors, including NGOs, IGOs and multinational corporations.

Further reading

For good introductions to globalization and global politics, I would suggest the excellent volume edited by J. Baylis and S. Smith, *The Globalization of World*

Politics (Oxford University Press, 1997) as well as C. Brown's *Understanding International Relations* (Macmillan, 1997). On critical geopolitics see G. Ó Tuathail's *Critical Geopolitics* (Routledge, 1996). For a flavour of realist and liberal institutionalist work on international politics, see H. Morgenthau's *Politics Among Nations* (Alfred Knopf, 1948), K. Waltz's *Theory of International Politics* (Addison-Wesley, 1979) and R. Keohane and J. Nye's *Power and Interdependence* (Harvard University Press, 1989).

Chapter 3

Global apartheid and North–South relations

The collapse of the Cold War has focused attention once more on the structure of the global political economy and the possibilities of changing divisions of wealth between North and South. It has been argued by many Third World and progressive writers in the North, such as Richard Falk, that the global political economy remains premised on a form of global *apartheid.* This presents a very different sense of globalization because it is premised on an assumption of inequality and difference rather than uniformity and mutual benefit. At this stage, the origins of the term 'apartheid' are worth considering.

In 1948 the South African government under President D.F. Malan introduced a set of policies and practices that became known in Afrikaans as 'apartheid' (separate development). Over the next forty years, elaborate plans were constructed not only to identify different racial groups (Whites, Blacks, Coloureds and Indians) but also to develop the South African economy and society along racial and ethnic lines. Politically, White South African citizens were the only category of people able to vote and participate in government. In terms of education, housing, social services and transport, strict segregation was enforced. Marriage between 'Black' and 'White' South Africans was forbidden, and residential areas were demarcated by racial classification. This system of apartheid was condemned by many countries in the United Nations because it actively, and often violently, suppressed the basic human rights of Black and other non-White peoples.

In 1990 the most famous prisoner in the world, the Black lawyer and activist Nelson Mandela, was released from detention in South Africa (Figure 3.1). Over the following years, the White minority government was forced to bow to domestic and international pressure to release scores of political prisoners, thereby beginning the process of dismantling apartheid as a prelude to constitutional change. In 1994 the first free and multi-ethnic elections were held in South Africa, with the result that Nelson Mandela became the first Black president of the country. However, in spite of the formal ending of apartheid, profound inequalities remain between White and Black South Africans. Under the leadership of Archbishop Desmond Tutu, the Truth and Reconciliation Commission (TRC), created in the aftermath of the 1994 elections, has attempted to expose the violent nature of apartheid to wider critical scrutiny.

We live in a world, as the American strategic thinker Thomas Schelling once noted, where one-fifth of the world is rich and predominantly

Figure 3.1 Nelson Mandela. (*Photo*: Jon S. Paull/Panos Pictures, © Jon S. Paull)

lighter-skinned and four-fifths are poor and darker-skinned. The richer peoples also enjoy an overwhelming military superiority and often seek to prevent the poorer folk from sending economic refugees to their developed regions (Schelling 1992, cited in Falk 1997). Military force combined with surveillance technologies continues to be used in order to prevent movement of 'economic refugees' from regions such as Latin America and North Africa to North America and Western Europe, respectively. Unsurprisingly, various international think-tanks have concluded that the unequal character of the global political economy had to be tackled:

> While most people of the North are affluent, most of the people in the South are poor; while the economies of the North are generally strong and resilient, those of the South are mostly weak and defenceless; while the contries in the North are, by and large, in control of their destinies, those of the South are very vulnerable to external factors lacking in functional sovereignty. . . . And the position is worsening, not improving. (South Commission 1990, cited in Falk 1997: 630)

A recent report by the World Bank, 'Everyone's Miracle?' (August 1997), also confirmed that more than two-thirds of the world's poorest people live in East Asia. In spite of the impressive growth rates of Asian tiger economies, absolute poverty and lack of educational opportunities have combined to ensure that millions of people in other East Asian countries such as Cambodia, Laos, Mongolia and China have to survive on less than one US dollar a day. Within this report, rural and agricultural communities in the East and South Asian region were perceived to be particularly vulnerable to abject poverty. India, for example, has at least 314 million people (1993–1994 estimate) living in extreme poverty. Non-governmental organizations have, however, often been critical of Northern-dominated institutions such as the World Bank because of their failure to address village-scale development and urban slum regeneration (Desai and Imrie 1998). In contrast, large-scale projects such as dam construction tended to dominate the funding profile of World Bank and other international agencies for the last fifty years. In the 1990s, World Bank and UNDP (United Nations Development Programme) reports on poverty and underdevelopment have tended to emphasize the significance of indigenous education spending and infrastructural investment without ever considering how North–South relations might impinge upon the capacity of the South to invest in these particular sectors.

This chapter is founded upon a belief that Northern debates over the new world order and the unequal impact of globalization have either neglected or marginalized the experiences of the South. The future of regions such as Africa, Asia and the Pacific in any new world order will depend upon the interaction of states coexisting within a globalized system of financial flows, social actors, militarization, markets, international organizations and unwanted ideas and threats. The position of countries in sub-Saharan Africa, such as Malawi and Uganda, is all the more precarious as it becomes evident that not even so-called great powers such as the United States can shape the international system to suit exclusively American needs. At a time of rapid political change, the inability to construct a new world order based on strong American leadership was clearly exposed in the face of economic weakness and domestic unease over spending commitments. This discussion of the South during the post-Cold War era concludes that the North–South cleavage can only be tackled by the progressive strengthening of a global civil society bolstered by an agenda of demilitarization (Chapter 5), cultural security, sustainable development, environmental protection (Chapter 6), human rights (Chapter 7) and global governance (R. Walker 1988; Falk 1995, 1997).

The Third World and the Cold War

The invention of the **Third World** by Western social scientists in the early 1950s coincided with the geographical extension of the systemic-ideological struggle between the two superpowers. It was perhaps no coincidence that new categories, such as First World and Third World, were being deployed at a time when the United States and the Soviet Union were directly involved in

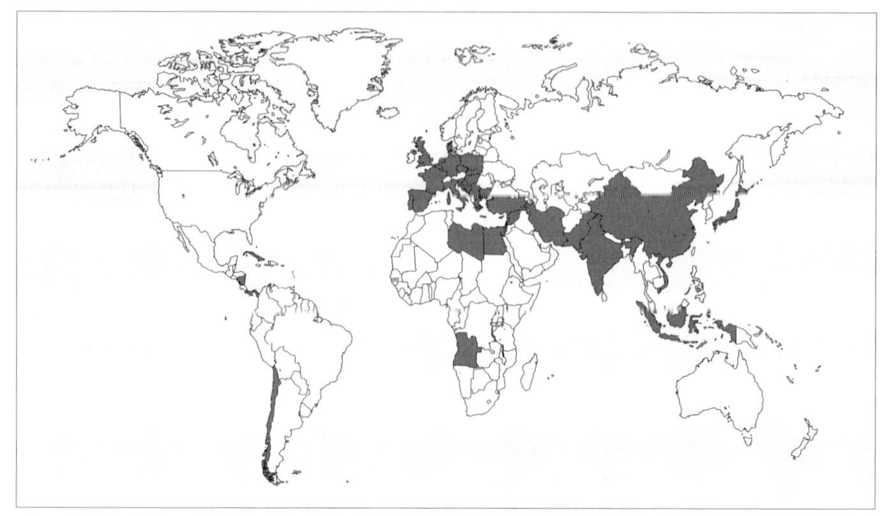

Figure 3.2 US–Soviet conflict: zones of most serious trouble, 1948–1988. (Adapted from Nijman 1992: 688)

supporting opposing sides in the Korean peninsula (Figure 3.2). Subsequent events in Korea, Vietnam and Central America were increasingly evaluated and judged within a narrative that stressed the significance of the ideological struggle between the superpowers. The geopolitical imagination of the Cold War was characterized by

> geopolitical space [being] conceptualised as a three-fold partition of the world that relied upon the old distinction between traditional and modern and a new one between ideological and free. Actual places became meaningful as they were slotted into these geopolitical categories, regardless of their particular qualities. (Agnew 1998: 111–12)

In the United States, successive administrations from Truman to Reagan adopted the geopolitical view that the Third World had to be saved from the enduring evils of communism and totalitarianism. In some cases this concern resulted in armed intervention in various parts of the world, ranging from the widespread carpet bombing of Cambodia in the 1970s to the despatch of twenty thousand marines to the Dominican Republic in 1965. Moreover, other countries such as Taiwan and South Korea received extensive financial and military assistance throughout the 1950s and 1960s because they were considered to be threatened by the Soviet Union. Taiwan, for example, derived 5–10 per cent of its national income from American financial aid in the 1950s (Ward 1997).

However, American commitments to the Third World were not geographically uniform. Throughout the Cold War, Latin America and the Caribbean were considered to be highly significant, whereas other regions such as West Africa were considered to be of lower geopolitical importance. This geographical variability has been noted in a recent analysis of the US president's State of the Union address from the 1940s to the 1980s (O'Loughlin and Grant 1990, cited in Agnew 1998: 116); see Figure 3.2. In the early stages of the Cold War, presidents tended to stress the threat to the so-called rimland states which surrounded the Soviet Union and China. In the 1960s, attention tended to be focused on the two socialist states of Cuba and Vietnam. By the 1980s, however, Presidents Carter and Reagan were expressing concern for the Middle East, Southern Africa and Central America.

While the overall pattern of concern may not be surprising, given the geopolitical contours of the Cold War, this analysis includes the consistently high priority given to Latin America and the Caribbean by American administrations. This concern for a neighbouring region was rarely benign, however. From 1945 onwards, American administrations developed a range of policies and strategies that were designed to protect Latin America from socialism and to promote American commercial and security interests. These included the creation of an inter-American security community (under the 1947 Rio pact), which involved mutual defence in the Americas, and the provision of financial and military assistance through programmes such as Alliance in Progress during the 1960s.

In more extreme cases, however, the American military and intelligence agencies were prepared to undermine governments in the Latin American region that were thought to be leaning towards the political left. In 1954 the Central Intelligence Agency (CIA) provided rebels in Guatemala with funds, arms and combat training so they could successfully overthrow the reformist government of Jacobo Arbenz Guzman (Immerman 1982). In 1961 the CIA also encouraged anti-Castro rebels to attempt an overthrow of the socialist regime of Fidel Castro. The subsequent failure of the so-called Bay of Pigs venture was not only a crushing indictment of the limitations of American power but also contributed to the worsening relations between the superpowers over Cuba. The decision by the Soviet Union to place missile installations on Cuba precipitated one of the most tense moments of the Cold War, when it appeared that the United States was prepared to launch military strikes against Cuba if the installation work continued. The crisis eventually ended when Soviet missile transporters were returned to their home bases.

In the same year as the Bay of Pigs fiasco, Third World states came together as a political force. The creation of the Non-Aligned Movement (NAM) in 1961 was an illustration of how some Third World states attempted to resist the international politics of the Cold War (Willetts 1978). The NAM was intended to be an organization of states such as India, Egypt and Yugoslavia which would resist the geopolitical pressures of the superpowers. Non-alignment is not the same as neutrality, because neutrality is usually a condition that is recognized or guaranteed by other states. Non-alignment is concerned with developing an independent political space which is secure from superpower interference. The founders of the Non-Aligned Movement in 1961 tried to create a political forum in which common problems, such as building a new state in the midst of the Cold War, could be discussed without interference from the two superpowers. Over the following twenty years, the NAM met at intervals of three to five years to consider the political and economic issues: Cairo 1964, Lusaka 1970, Algiers 1973, Colombo 1976, Havana 1979, New Dehli 1983, Harare 1986, Belgrade 1989, Bogota 1994. Although it had no central headquarters, the NAM did coordinate activities on technical cooperation, development, disarmament and international security. Summit meetings were the major venues for debate and policy formulation (Singham and Hune 1986).

At the 1973 NAM summit, the parties committed themselves to pursuing a new international economic order (NIEO) to reduce the North–South divide. This NAM summit in conjunction with the rise of oil prices by OPEC (Organization of Petroleum Exporting Countries) in 1973 prompted discussion of the NIEO at the United Nations in 1974. Despite the high profile of the NIEO debates, the NAM never really enjoyed high-level political success, because its members were divided on the ultimate objectives of non-alignment. Some countries such as Cuba and Libya wanted the NAM to align itself more closely with the Eastern bloc, whereas others argued that the movement should look to the West for political support. By the late 1970s, arguments for an NIEO had declined in political salience, not least because the reemergence of a Second Cold War had shifted the political agenda away from economic issues. With the ending of the Cold War in the late 1980s, the political significance of the NAM had largely disappeared. But new members such as South Africa (1994) have ensured the organization continues to meet in order to discuss the politics of non-alignment in the 1990s.

These struggles for survival should not be underestimated, given the scale and intensity of violence in many parts of the Third World. In Southeast Asia, for instance, over 600,000 local people died due to the confrontations between rival American and Soviet-backed military forces between 1969 and 1975. In other parts of the world, socialist and military regimes in Africa, Latin America and Asia strove to consolidate the powers of the state within a rapidly changing world economy. Socialist governments such as Mozambique and Angola were wrecked by civil wars and superpower intervention (in Southern Africa) in the 1970s. Over one million people are believed to have died between 1975 and the early 1990s in Mozambique alone (Sidaway and Simon 1993). International agencies such as the World Bank had to provide emergency financial aid in order to save these states from total collapse due to civil war, and the early achievements in health care and education provision were destroyed.

The Non-Aligned Movement succeeded in changing the often violent profile of North–South relations through its adoption of a campaign for a new international economic order (NIEO) based on financial and technological transfers from North to South and through the promotion of peaceful cooperation between states (Thomas 1987; Halliday 1989). The initial impetus for an NIEO stemmed from the development at the United Nations Conference on Trade and Development (UNCTAD) and the creation of the Group of 77 within the United Nations in 1964. The Group of 77 represented the poorest member states of the United Nations and was designed to bring Southern voting power to bear on Northern member states of the UN Security Council. The meetings of the UN General Assembly and UNCTAD were used to raise the issue of unequal trading relations between North and South. The demands for an NIEO were based on a belief that radical change was needed in order to improve the condition of the South. Basic demands included a new general system of preferences to enable the South to break into the manufacturing markets dominated by the North; a commitment from the North to devote at least one per cent of GDP to official aid; the cancellation of the 'Southern' debt; technology transfers to be executed; and the improvement of control and regulation of multinationals to prevent the exploitation of Southern resources and labour markets.

This was an ambitious agenda which demanded radical reforms of the international economic order. It was also conservative in the sense that co-operation between states was still considered to be the best means of promoting economic development for the South within the capitalist world economy. However, it was also grounded on a belief that structural obstacles within the global political economy would have to be overturned. In the late 1970s there appeared to be some evidence that the South was making progress and that even the UN-appointed Brandt Commission (named after the former German chancellor Willy Brandt) recognized the significance of these inequalities between North and South. Furthermore, the South proved to be an effective negotiating bloc during the oil price rises of 1973–1974 and the United Nations Convention on the Law of the Sea in the 1970s and 1980s. The declaration of the ocean floors as common heritage (and therefore the property of the global community) was a considerable political success, despite American and Northern opposition. However, fundamental change in the world economy was elusive in the 1980s, as priorities changed and the onset of the Second Cold War ensured that Northern states were more concerned with rising superpower tension than North–South relations; Figure 3.3 indicates the massive US defence spending. By the time of the 1982 World Summit of Northern and Southern Leaders in Mexico, it was abundantly clear that Northern leaders such as President Reagan and Prime Minister Thatcher had no interest in meeting the demands of the NIEO.

The Northern states' apparent lack of interest in fundamental reform led Southern states and their commentators to talk of a so-called lost decade of development (Green 1995). Throughout the 1980s, the political and economic condition of many parts of the Third World began to worsen as economies collapsed in sub-Saharan Africa, and Central America witnessed the destabilization of Nicaragua and the invasion of Panama. The renewed geopolitical confrontation between the Soviet Union and the United States therefore had dire consequences for the economic and political welfare of the Third World. Armed intervention combined with rising debt burdens and public service sector collapse prompted discussions of so-called failed states, a term first introduced in the 1980s to convey a sense of place where the basic mechanisms of governance had simply evaporated. For Mozambique,

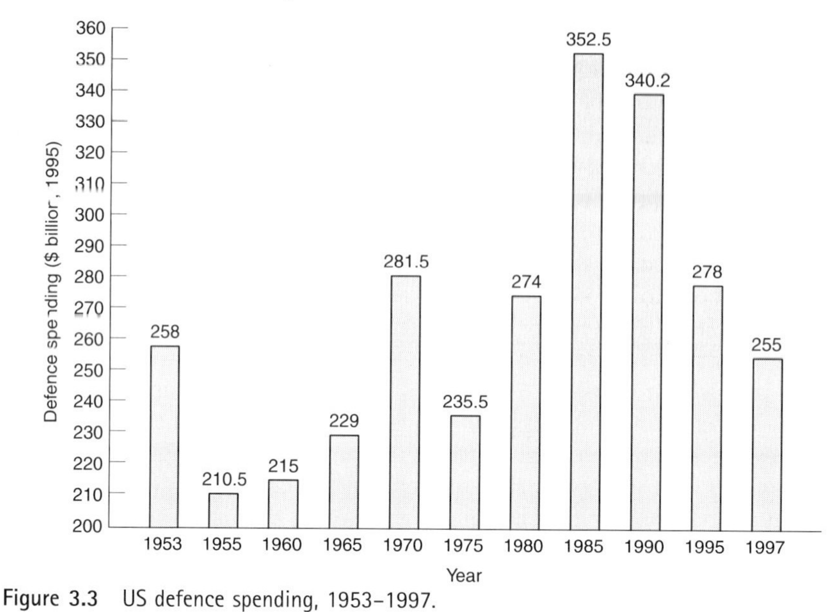

Figure 3.3 US defence spending, 1953–1997.

governance was increasingly determined by international bodies based in Washington DC rather than in the national capital, Maputo.

By the end of the Cold War, NAM and the demands of the NIEO had lost their economic and political appeal because of the changing relationships between NAM's members, the superpowers and the wider international community. The onset of the debt crisis in 1982 (see below) further compounded the South's inability to demand fundamental change in spite of the initial shock to the Northern financial community. Within the Southern coalition, the collective demands for radical reform were also beginning to fragment as it became apparent that some states such as South Korea and Malaysia had enjoyed considerable success in terms of economic growth and rates of industrialization. For world systems theorists, the growth of a Southern semi-periphery was a natural outcome in the sense that the world economy needed economic and political safety valves. It was therefore in the North's interests that some Southern countries developed successfully whereas others remained underdeveloped. The rapid political changes of the 1980s induced some analysts and political leaders to argue that the South, or the Third World, had effectively ended because of the diversity of experience in the regions. New times therefore demanded new political programmes and new forms of analysis.

The end of the Third World?

Since the end of the Cold War and the collapse of the Soviet Union, increased attention has been paid to the intellectual utility of Cold War categories such as First World and Third World. It has been widely suggested that the term

'Third World' is no longer an appropriate label for the complex and varied regions of North Africa, South Asia, sub-Saharan Africa, Latin America and the Caribbean, Southeast Asia, Southwest Asia and the Pacific (Berger 1994; Ayoob 1995; Grant 1995; Haynes 1996). During the 1990s, critical observers in the North and South advanced three major objections to the concept of a Third World. The first could be described as a philosophical objection to the implicit assumption of three different worlds. The concept of a Third World erroneously assumed that the lives of human beings in Africa, Asia and Latin America were entirely separate from those living in the First and Second Worlds. As globalization theorists have stressed, all human beings live in one, and only one, interdependent world. The formation of an industrialized North and an underdeveloped South was intimately related rather than derived from separate economic and political processes. Moreover, the differentiation between First, Second and Third Worlds implicitly assumes a value hierarchy where the First World is considered superior to the Third World.

During the Cold War, the term 'Third World' had an apparent analytical value because it seemed to refer to states who not only shared a common colonial experience but were also intent on economic development. Mainstream development approaches in the United States ensured this categorization also implied that the Third World should be seeking to follow the example of the First World. Walter Rostow's manifesto for a non-communist approach to economic development assumed there were five major stages of development which would involve a substantial transformation in the cultural, economic and political life of developing nations. It was generally assumed that development would be a relatively uniform process for the Third World states, regardless of their particular location and history. The division of the world into three separate spheres meant in practice that Western observers tended to neglect the interrelationships between these allegedly separate worlds.

The second point of objection is concerned with the ending of the Cold War. The concept of the Third World was developed in the 1950s by Northern social scientists to refer to a world dominated by the bloc politics of the Cold War. A tripartite division of the world made some sense in the 1960s, when the world was characterized by a superpower confrontation and the emergence of newly independent nations in Africa and Asia. However, these circumstances changed radically, and in alliance with the acceleration of political and economic globalization, the world has witnessed the rapid transformation of the earth's political geography. Some parts of the Third World have become highly developed whereas others have floundered. The so-called East Asian tigers have experienced some of the highest economic growth rates in the last twenty years. The collapse of Second World federations such as the Soviet Union and Yugoslavia has meant that some of the former Soviet republics such as Armenia are alleged to resemble Third World economies. More generally, a shift of geoeconomic influence from the Euro-American realm towards the Asia-Pacific basin has meant that the political geography of the post-Cold War era is quite different from that of the 1950s and 1960s.

One major element of change in the political geography of the world economy has been the rising profile of China. China has been described as the next economic and political superpower after the United States and Japan. As early as 1975, the *Economist* magazine was predicting that China's expanding economy would be a major force in the world economy. To date, China's economy has grown annually at around 10 per cent since 1991 and it now produces half the world's toys, two-thirds of its shoes and most of the globe's bicycles and power tools. China is also the largest recipient of foreign investment after the United States. In 1995–1996 economists estimated that China's gross national product (GNP) would exceed both the United States' and Japan's by the middle of the next century. Under the leadership of Deng Xiaoping, China engaged in a massive programme of market reform and commerical development in the 1980s and 1990s. There is little doubt that standards of living have improved for many Chinese people in terms of access to clean water, possession of consumer goods, and better food and housing. However, the environmental and social costs have been high in terms of poor employment conditions for many workers, water shortages, environmental degradation due to industrial pollution and continued controversies over the state of human rights in the country.

The final point of objection to the term 'Third World' belies the condition of elites within these states. The promotion of Third Worldism in the 1970s and 1980s often disguised the fact that Third World elites (often Western-educated) were not always acting in the best interests of their own societies. Notorious political leaders such as President Idi Amin of Uganda stole millions of pounds and dollars from the Ugandan government and deposited the money in secret Swiss bank accounts. He later attempted to kill or expel all the ethnic Asian Ugandans in the early 1970s in a bid to ethnically cleanse Uganda of 'foreign' elements. In spite of its rich natural resources and exports such as coffee, Uganda is now one of the most heavily indebted countries in the world. In the Central African Republic, the former self-styled 'Emperor' Bokassa spent US$20 million (equivalent to 25 per cent of the total GDP) on his coronation ceremony in 1977. In Zaire, former President Mobutu stole several billion dollars over a period of twenty years, money derived from the country's export in oil and diamonds (Reyntiens 1995). Categories such as the Third World effectively homogenized conditions within these parts of the world rather than exposing the enduring and contradictory complexities of these post-colonial societies. Within the socialist world of Third World states, appeals to equality and socialist forms of development were often overwhelmed by high levels of violence directed against an internal population. The 'killing fields' of Kampuchea (now Cambodia) in the 1970s are a chilling reminder of how a socialist regime led by the Khmer Rouge leader Pol Pot participated in the massacre of one million people (Figure 3.4).

Far from ushering in a new global order based on uniform economic development and liberal democracy, the conditions of the Third World remain so varied that the standard social science categories such as 'developing countries' and the 'periphery' increasingly do not make sense for countries

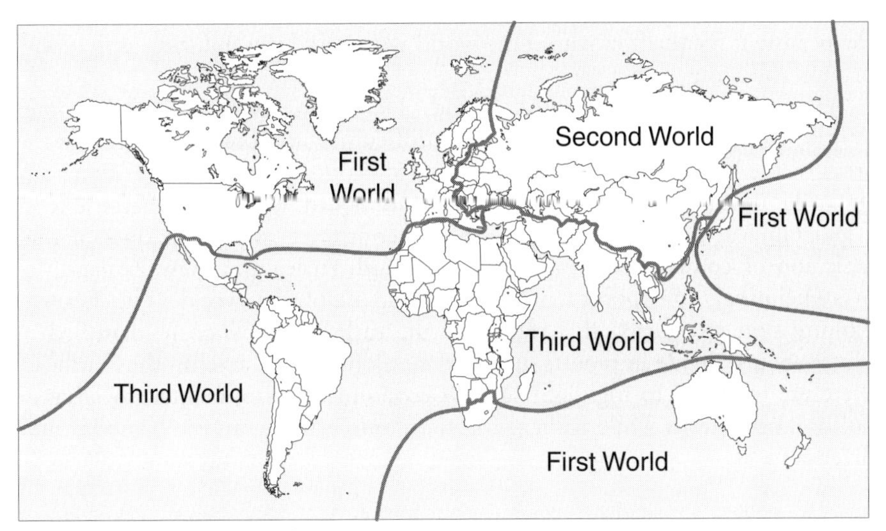

Figure 3.4 The three worlds?

ranging from Cambodia to Yemen and from Singapore to Togo. Robert Gilpin noted in 1987 that the Third World 'no longer exists as a meaningful entity' (Gilpin 1987: 304). Rapid political change has therefore apparently called into question the capacity of mainstream concepts and theories to explain and interpret the world around us. As Cedric Grant has claimed:

> Since the collapse of communism and the end of the Cold War, the scepticism as to the existence of the Third World has increased. This is because the term 'Third World' was derived in the context of a bipolar world as a label to differentiate the newly independent countries of Africa and Asia from the rival power blocs, the Western and the Soviet, which in their competition with each other had focused their attention on these newly independent nations. Even those who were inclined to agree that there was some substance to the concept of the Third World are now more ready to accept the contention that the global transformation which is occurring is rendering the concept anachronistic. (Grant 1995: 567–68)

The actual delimitation of a 'Third World' during the Cold War deserves further elaboration because it touches on some of the enduring controversies surrounding those countries in Africa, Asia and Latin America which have yet to achieve economic wealth comparable with states in Western Europe and North America, and also Japan. This discussion will broach the intellectual and academic context that gave rise to the concept of a Third World. The purpose of this investigation is to demonstrate that the conceptual challenges posed by the concept of a Third World are far greater than a simple presentation of the changing political map in the post-Cold War era. In an era of increasing globalization, the advocacy of a concept such as the Third World could be used to promote a spurious impression of homogeneity, thereby reproducing an unhelpful distinction between a First World and a Third World. On the other hand, the term 'Third World' can be useful in highlighting the persistent inequalities within the world system and the enduring aspirations of several billion people.

As with any label – Third World, Developing World, Low Income World, etc. – there are always inherent difficulties in representing either vast areas of the earth's surface or complex socio-economic situations in terms of single categories (Barton 1997: 6). The term 'South' is preferred because it is a geographical reference to the southern hemisphere, which, in spite of the inclusion of countries such as Australia, South Africa and New Zealand, is overwhelmingly the poorer hemispheric region of the world. The Brandt Commission recognized this feature in the early 1980s when it identified a North–South divide in reports on world development. The term 'the South' is therefore intended to highlight similiar economic, environmental, social and political conditions while recognizing that Southern regions are complex and diverse.

US–Latin American relations

The British geographer Doreen Massey employed the term 'power-geometry' to highlight the unequal and paradoxical nature of globalization. On the one hand, Northern governments and social commentators frequently depict the earth as a 'borderless world' composed of unfettered spaces; and on the other hand, they also seek to control and regulate movement and flows within bounded spaces. The immigration controversies in the United States and Western Europe reveal the desire of rich countries to restrain the movement of poorer peoples while simultaneously demanding the free movement of capital and investment. In California during the 1990s, voters were debating proposition 187, which set out to prevent illegal immigrants from accessing any form of public service such as health and welfare. Yet, at the same time, these illegal immigrants provided services such as office cleaning, which the local populace was unwilling to do, given the pay and conditions. These spatial inequalities ensure that the poorer regions of the world are held in place and invaded by the rich in terms of economic investment and political interference. For poorer regions such as Latin America, the ending of the Cold War has not radically changed the political economic condition of the population. As a Mexican political scientist has noted:

> Latin America . . . finds itself in a sadly paradoxical bind. The end of the Cold War has brought greatly broadened geopolitical leeway, but economic globalization and ideological uniformity have rendered that at least partially meaningless. (Castaneda 1994: 48)

Investigating the role of the South in the post-Cold War era is a necessary component for any critical evaluation of globalization. The South, as Jonathan Barton has argued, cannot be considered peripheral to such an investigation (Barton 1997). In Nicaragua, a country caught up in the ideological and territorial struggles of the Cold War, per capita income has fallen in real terms since the 1960s as a result of economic pressure from the North, geopolitical destabilization by rebel forces and US military support of antigovernment forces. In Guatemala, where two per cent of the population own 63 per cent of the most productive land, the ending of the Cold War did not lead to a transformation of land ownership. Moreover, the US invasion of Panama in 1989 reminded Central Americans that the sole remaining superpower is not averse to violent intervention in the region. The removal of the country's leader, General Noriega, was ironic given the US administration's previous support. Other commentators have also pointed to the fact that the United States was concerned about the growing levels of Japanese investment in the Panamanian Isthmus and was thus anxious to restore its geopolitical authority over the area. The Panama invasion was a significant development, as it was the first hostile post-Cold War incursion. As the Honduran newspaper *La Tiempo* noted in December 1989:

> It was a coarse grotesque euphemism [Operation Just Cause: the code name for the American invasion], neither more nor less than an imperialist invasion of Panama. . . . We live in a climate of aggression and disrespect . . . hurt by our poverty, our weakness, our naked dependence, the absolute submission of our feeble nations to the service of an implacable superpower. Latin America is in pain. (Cited in Chomsky 1991: 158)

The invasion of Panama coupled with massive destabilization of Central American governments by the superpowers contributed to the so-called 'lost decade' of development and social progress in the 1980s.

The failure to eradicate the debt burden of the South is probably the single most enduring inequality between North and South. In 1990 it was estimated that the total debt of the South/Third World had reached US$1.4 trillion. In Latin America the debt burden accounted for a substantial amount of total export earnings: Mexico US$85 billion, Brazil US$105 billion and Argentina US$61 billion (1993 figures). The most indebted continental region remains sub-Saharan Africa when measured by total external debt in relation to the export of goods and services (Simon et al. 1995). Through a combination of factors, including the rapid rise of lending by Northern banks and states in the late 1970s, Southern states accumulated substantial debts by the 1980s because of their incapacity to repay loans and grants. Global economic depression in the 1980s further contributed to this so-called lost decade of development for Latin America and sub-Saharan Africa. The suspension of debt repayment by Mexico in August 1982 precipitated the biggest financial crisis in the history of the international financial system. Shortly afterwards, other states such as Brazil and Argentina suspended their debt repayment schedules too.

Over a period of fifteen years, the international community has promoted a range of debt-rescheduling packages for countries such as Mexico. With the assistance of the United States, the Mexican government was instructed by the World Bank to follow an austerity package, which sought to devalue the national currency and cut public spending, in order to reduce the annual burdens on the Mexican treasury. However, after a decade of financial austerity the country was hit by further financial crises, which led to the collapse of the Peso, the withdrawal of foreign investment and a decline in economic growth. In 1996 the Mexican debt was estimated to be around US$85 billion, at a time when hopes for a new deal with the World Bank and the IMF had been envisaged.

The recent experiences of Mexico have been repeated, admittedly in different ways, around the countries of the South. Attempts to structurally adjust debt-ridden economies have not been successful in promoting sustainable development or reducing poverty and hunger in the South. The idea of structural adjustment policies was to liberate extra monies for debt repayment through public sector reductions in spending. This has not been effective in terms of building a more sustainable future for Southern societies because economic plans tended to emphasize reductions in consumptions rather than investment for people in the future. In Latin America, the United States has been actively involved in reducing debt levels because of the geographical and political economic proximity of countries such as Mexico. It has been argued, for instance, that American plans to create a North American Free Trade Association (NAFTA) depended among other things on Mexico's financial position being improved by the 1980s. Debt-relief plans for Mexico were implemented by the Reagan administration to increase confidence in the Mexican economy. President Carlos Salinas of Mexico later claimed that an 'economic miracle' had occurred between 1988 and 1994 because of the rise in foreign investment in the form of speculative capital.

The subsequent financial crisis in Mexico in the mid-1990s sparked off a wave of protests against structural adjustment and debt burden. In January 1994 a guerilla uprising by the Zapatista National Liberation Army (EZLN) in the southern states coincided with Mexico's formal entry into NAFTA. Drawing their inspiration from the Mexican revolutionary leader Emiliano Zapata, the Zapatistas (EZLN) captured the town of San Cristobal de las Casas in the Mexican state of Chiapas on 1 January 1994. Under the leadership of Subcommandante Marcos, the Zapatistas declared war on the Mexican state and demanded, among other things, that indigenous cultures and basic human rights must be respected. Through their skillful use of the media, including the Internet, the Zapatistas were able to mobilize considerable support for their campaign from NGOs and human rights campaigners in Mexico and in the wider world. In February 1995, President Zedillo ordered the Mexican armed forces to crush the rebellion after pressure from those international investors who were concerned that the Mexican economy and society were out of control.

Although the struggle for land rights and the protection of local cultures continues, the Zapatistas demonstrated that Third World social movements can use media sources to raise the profile of particular campaigns and garner international support for their causes. An uneasy stand-off exists between the Zapatistas and the Mexican government. For critical geopolitics, these activities demonstrate that realist views of political life fail to capture the ways in which state-centred notions of hegemony, power and territory may be challenged (Routledge 1998). These events reaffirmed Mexico's dependency on

American financial and political cooperation, and prompted the Mexican economist Carlos Heredia to make this appeal:

> We need a policy mix that will identify key sectors requiring protection over time. NAFTA and the bail-out package [US package negotiated in 1995] are an obstacle to that, as they make it impossible to regulate capital and investment flows. . . . We need an economic strategy which stimulates production at the small, local and medium level and puts producers, workers and society on the centre stage. But first, Mexico's debt payments *must* be suspended. (Original emphasis; cited in Latin American Outlook 1995: 1)

This view was reiterated by Latin American observers in the wake of disasterous floods that cost many lives and wrecked the infrastructure of Honduras, Guatemala and Nicaragua in October 1998.

Southern views on development

For the last fifty years, official development policies have tried to promote development through the political and economic transformation of states in the South (Escobar 1995; Rist 1997). It could be argued that, by any conventional indicator of development, these policies have failed. In 1997 it was recorded that in nineteen countries per capita income had fallen below the 1960 figure. Poverty and hunger continue to affect vast areas of the world including ethnic minorities, the disabled and the elderly in the North. Over one billion people still do not have access to clean water supplies and it has been estimated that, in terms of global income distribution, 82 per cent of total income is owned or enjoyed by the richest 20 per cent of the global population (United Nations 1998). In that sense, World Bank figures for gross domestic product (which do not consider patterns of distribution) tell us little about the lives of people living in slums, nor do they remind us that far more people have died from disease and hunger than the 187 million people who perished through wars and conflict in the present century (Hobsbawm 1997).

There is a lengthy, if neglected, tradition concerned with the actual conditions of the South within the global political economy. 'Southern' views of international politics have been constructed on a more general account of the centre–periphery relationship within the world economy. These accounts are 'Southern' in the sense that the writers hail from Latin America, Africa and Asia rather than the Euro-American world. In the 1950s, for example, the economic writer Raul Prebisch, an Argentine economist working at the United Nations Economic Commission for Latin America, proposed that the industrialization of the South was being restrained by the North and the workings of the capitalist world economy. He argued that the South's dependence on the production of primary products for the North, coupled with the consumption of goods manufactured in the North, was inherently disadvantageous to the South. In the long term, trading conditions force the South to derive ever more credit from primary exports in order to retain purchasing power. Unlike manufactured goods and services, primary products do not provide much scope for innovation and increased profitability. Many Southern states therefore have little alternative other than to retain the economic and political position in a Northern-dominated international economic order.

In the 1960s new writers such as A.G. Frank and F. Cardoso (now the president of Brazil) directed the focus of analysis towards class relations and patterns of exploitation. One of the key areas of debate was the extent to which Southern capitalists and governments were junior partners in a global system of exploitation and domination. In his path-breaking analysis *Capitalism and Under-Development in Latin America* (1971), Gunders Frank presented a detailed argument of the systematic underdevelopment of the South. In essence, Frank claimed that the promise of economic development for the South was not only inherently false but also that the South was actually participating in its own underdevelopment. The structural constraints on the South were such that economic development was always likely to be minimal and precarious because of the Northern domination of the world economic order. Though later criticized for their economic and political assumptions about class, the state and the world economy, these kinds of ideas were emblematic of a wider concern for the condition of the South. The demands for a New International Economic Order (NIEO) in the 1970s could be attributed to the work of structuralists such as Frank and Cardoso.

Although these accounts of the global political economy have been critiqued over the years, the dependency writings contributed to a rather different series of perspectives on international relations. For much of the post-war period, the disciplines of geopolitics and international relations have been resolutely Anglo-American in the sense that most of the Northern-based writers were concerned with either the North and/or the international system *per se*. Following from this body of literature, **world systems theorists** such as Immanuel Wallerstein and Peter Taylor argued that social and political relations between the North and South need to be considered within a longer time frame of an evolving capitalist world economy (Wallerstein 1980; P.J. Taylor 1993). So the conditions of the South in the 1990s have to be investigated as part of a longer historical process. A range of governments in Africa and Asia have tried, as post-colonial states, to secure vulnerable national territories and economies in the face of weak state sovereignty. During the Cold War, for example, many nations of the Third World experienced direct interference and military intervention from outside powers in order to undermine a particular regime. The human cost of these interventions was very high, as nations such as Mozambique and Angola were destabilized with dire consequences for civilians, particularly women and children.

Northern debates over globalization have been intensely concerned with the erosion of state sovereignty and transboundary political and economic flows. In the South, experiences of this kind have been routine in terms of the undermining of state jurisdiction and the penetration of Western influences into national cultures. Mohammed Ayoob and Caroline Thomas have argued that the economic dimensions of national security, such as access to secure systems of food, health, money and trade, are major concerns for Southern states (Thomas 1987; Ayoob 1995). No wonder then that governments of the South have often been staunch supporters of the principle of non-intervention, mindful of the fact that the international system is not based on the premise of equal and self-determining sovereign states (Chapter 7). States such as the United States have been far better equipped to deal with the demands of international politics and globalization, whereas others such as Sudan and Mozambique might be best described as quasi states in the sense that their continued existence and legitimacy has more often than not been derived from international relations rather than internal support. Recent debates over human rights, societal security and humanitarian intervention in the 1990s had substantial implications for the South and its capacity to protect a futher erosion of the right to conduct its own affairs.

During the 1990s it has became apparent that a number of pressing issues confronting the South and South–North relations have not been resolved: the political and economic consequences of development, gender and human rights, environmental protection, debt reduction and the protection of ethnic and religious minorities (Haynes 1996). At the same time, mainstream development approaches have failed to tackle the underlying structural causes of poverty, hunger, disease and chronic indebtedness. Major international conferences such as the 1992 Rio summit and the 1995 Conference on Socio-economic Development have tended to reaffirm a commitment to the promotion of free trade, market integration and liberal democratic governance, but for 'Southern' critics and NGOs these platitudes do not confront the profound inequalities of the global political economic system. Moreover, Southern critics have expressed anger at those Northern critics who blame Southern population increase for global enviromental change, rather than acknowledging the massive consumption of raw materials by the North.

In contrast, attention in the South has focused on promoting local forms of development, which stress local needs, self-reliance, ecological sustainability and community survival. Southern NGOs, in alliance with Northern NGOs and progressive commentators, have called for new forms of development strategies. Local groups such as the Chipko movement in India and the rubber tappers' movement in Brazil have been lauded for their campaigns to protect access to their environments and resources. Other groups in Guatemala and Ecuador have highlighted the importance of land reform in these countries, where the vast majority have no means of growing their own crops and developing sustainable lifestyles. For the poor of the South, sustainable development is a fiction when rich minorities control most of the fertile agricultural land, leaving the poor in places such as Brazil (where one per cent of the population owns 48 per cent of the land) and Zimbabwe (where two per cent claim 60 per cent of the land) to exploit fragile uplands and/or rainforests in order to meet their needs.

Conclusion

In the South, the recent transition towards market-based economies and liberal democracies has often been fraught. For one of the poorest countries in the world, Mozambique, the transition from a socialist developmental project to capitalism has been deeply problematic given the state of the country after twenty years of civil war and external intervention. Mozambique's economic and political condition remains parlous, even with the ending of the civil war in the early 1990s and recent elections. In spite of Mozambique's improvement in GDP growth, the destruction of basic education and health provision provides a grim reminder of the profound differences between North and South. Although forms of entrepreneurship and private sector growth occur in Maputo, the majority of the population remains impoverished and unwanted even by neighbouring South Africa, which constructed an electrified boundary fence in order to prevent illegal migration from the state. Another example, therefore, of a state trying to strengthen its territorial boundaries in the face of apparently unregulated flows of refugees and migrants.

In terms of globalization and geopolitics, this chapter on North–South relations has warned against simplistic assumptions about a world divided (in the form of global apartheid) into an improverished South and a rich North. The architecture of division (within the world) is more complex, as some parts of the North are as improverished and socially excluded as the South. The mortality rates for Black American children in the southern part of the United States are as horrendous as in many parts of the South. Likewise, some of the elites found in Southern cities such as Mumbai and São Paulo would compare favourably with their Northern counterparts in London, New York and Tokyo regarding access to consumer goods and lifestyles. However, these words of caution should not disguise the fact that profound economic and political divisions between North and South will persist well into the next century, notwithstanding changes in particular countries and economies such as the East Asian tigers. For some sceptical commentators, the prospects for the Third World appear bleak because of three major factors: a reduction in aid and investment from the North to the South, a rise in racism and anti-immigration politics in the North, and an increased tendency by powerful states to pressurize the South over debt rescheduling and trade access. For many commentators in the South, the current penchant for securing 'market access' to the world economy will ensure that Northern states continue to exploit the vulnernable and poorer zones of the world economy. Although the rationale for the Cold War may have disappeared, the forces of economic globalization and supranational capitialism will ensure that the power-geometries of North–South relations will remain unequal and fractured.

Further reading

For very good summaries of North–South relations and the Cold War, see F. Halliday's *Cold War, Third World* (Verso, 1989), C. Thomas's *In Search of Security: The Third World in International Relations* (Harvester, 1987). On non-alignment see P. Willetts' *The Non-Aligned Movement* (Pinter 1978) and A. Singham and S. Hune's *Non-Alignment in an Age of Alignment* (Zed, 1986). On development see A. Escobar's *Encountering Development* (Princeton University Press, 1995), G. Rist's *History of Development* (Zed, 1997), and 'Rethinking Geographies of Development', *Third World Quarterly* **19** (4) 1998, a special issue edited by D. Simon and K. Dodds.

Chapter 4

Popular geopolitics

Most people in North America, Japan and Europe learn about foreign affairs through the media, whether it be via the television, the radio or the Internet. It is probably not unreasonable to assume that television coverage played a key role in raising public awareness of the humanitarian crises in places such as Kosovo and Rwanda. Over the last thirty years, television and other communication technologies have contributed greatly by collecting, presenting and circulating information about the world (Figure 4.1). According to some observers, these technologies have enabled distant events and places to be transported via images and news stories into the living rooms of people in the North. The transmission, circulation and reception of information is not a neutral process, however. Some places, such as Bosnia and Iraq, have received considerably more television coverage in the 1990s compared with the civil wars and humanitarian emergencies in Angola, Kashmir and the Chechen province in Russia, which has led many media observers to conclude that television coverage is inclined to shape the political agendas of the United Nations and major powers such as the United States, France, Russia and the United Kingdom.

This chapter introduces the possibility that images and representations of world politics can be tremendously important in shaping patterns and responses to world political events. Not all television stories, newspaper features or films are simply propaganda, hence it will be useful to investigate how various sources construct particular interpretations of events, places or processes such as the Cold War, interpretations which influence courses of action. For our purposes, propaganda is defined as the deliberate construction and release of information designed to mislead or misrepresent particular situations and circumstances.

Traditional geopolitics has tended to assume that the geographical assumptions, designations and understandings of world politics are restricted either to the formal geopolitical models of Halford Mackinder or the policy statements of national leaders or Secretaries of State. The term 'popular geopolitics' is used to explore how societies and states often attempt to represent the world and their position in consistent and regular ways (Ó Tuathail 1996); see Figure 4.2. Formal architecture, such as monuments, and media sources, such as television, music, film, magazines and cartoons, provide resources and/or even actively construct particular vistas of world politics and specific places. In contrast to many mainstream realist and liberal approaches to world politics (formal geopolitics), critical geopolitical authors have argued that ideas

Figure 4.1 News network CNN became renowned for breaking stories before its rivals.

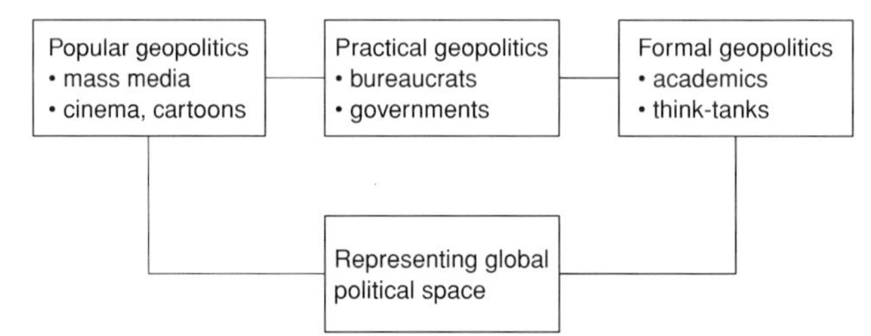

Figure 4.2 Linking popular, practical and formal geopolitics. (Adapted from Ó Tuathail and Dalby 1998: 5)

and representations about the political world are expressed and reproduced outside the narrow confines of the diplomatic circuit, foreign policy decision-making and intergovernmental conferences (practical geopolitics).

In Argentina, for example, stories and representations of the Islas Malvinas (Falkland Islands) are to be found on murals, postage stamps, atlases and

Figure 4.3 Monument to the war dead of the 1982 Falklands/Malvinas War. The inscription along the plinth salutes the Argentine soldiers who gave their lives for the attempted liberation of the Malvinas Islands. (*Photo*: Klaus Dodds)

monuments, in popular songs, countless newspaper articles and television shows (Figure 4.3). Underlying these representations remains a belief that Britain illegally colonized these Argentine islands in the Southwest Atlantic. Since the 1830s, successive Argentine governments have not only protested against this occupation but have also sought to remind their citizens that Argentina remains a geographically incomplete nation until the reclamation of the Islas Malvinas. By raising public awareness, it becomes perhaps more understandable how and why an unpopular Argentine military regime could gather substantial popular support for the invasion of the Falklands in April 1982. It is hoped that the first official visit of President Menem of Argentina (the first by an Argentine president since 1961) to the United Kingdom in October 1998 will lead to more cordial relations in the future.

As established in earlier chapters, geopolitics is considered to be a series of problematics concerning power, knowledge, space and identity. We have already explored, for example, how geopolitics can be considered as a particular discourse on statecraft and state power, and how and with what consequences particular networks of power-knowledge construct hegemonic geographical and political identities. An examination of popular sources such as television and magazines offers critical geopolitics ample possibilities for discerning how other places and peoples are represented within a variety of national and cultural contexts. Five such popular sources are considered: films, television, the popular magazine *Reader's Digest*, cartoons and music.

Films and geopolitical visions

The film industry has played a very important role in the twentieth century. As a technology it is ideally suited for projecting political, social, moral and cultural views to audiences around the world. American cinema has had a long and complex relationship with political institutions, public opinion and national identity formation. One of the earliest films, D.W. Griffith's *The Birth of a Nation* (1915), which blamed Afro-Americans and devious politicans for the American Civil War, was thus highly significant in constructing particular narratives about American national identity. As Ó Tuathail noted, '*The Birth of a Nation* portrays the Ku Klux Klan as the saviours of the white race, as defenders of the virtue of white women, and as representatives of the Christian civilisation, a civilisation under threat from the innate primitism and uncontrollable sexual appetite of emancipated African-Americans' (Ó Tuathail 1994: 540).

These early silent cinematic productions were very popular with American and other European audiences. In the enclosed atmosphere of the film house, cinema had a tremendous potential to hold the attention of the audience. Novel forms of visual presentation were quickly adopted by political and media elites alike, as a powerful propaganda tool which could be used to screen epic tales of nation formation and identity politics. In the Union, the Soviet Communist Party funded cinematic projects such as *October* (1927) followed by a stream of films in the 1930s such as the *Two Captains* and documentaries depicting the Arctic exploits of Soviet pilots (McCannon 1998). The 'myth of the Arctic' became a central feature of Stalinist popular culture as Soviet citizens were cajoled into taking an interest in the exploration and 'conquest' of the North Pole. The purpose of such a financial and cultural investment was to demonstrate that the communist state could conquer any obstacle placed in front of it, natural or unnatural.

In the United States it was often noted that wars tended to be fought twice: first on the battlefield and then on celluloid. The relationship between governments and film production companies was often intimate, given the enormous potential for influencing public opinion. In the first half of the twentieth century it was common for governments, including most famously Hitler's Germany, to cooperate very closely with film-makers on specific projects. Over the last eighty years there have been countless examples of the American government cooperating closely with Hollywood. The film *The Battle Cry of Peace* (1915) witnessed the former president Theodore Roosevelt asking the then Wilson administration to approve the release of 2,500 marines for the picture, which concentrated on exposing the fragile nature of America's armed forces at the beginning of the First World War. It was mooted that the popularity of this film helped to persuade the American president Woodrow Wilson not only to build up America's fighting capacity but also to enter the First World War in 1917.

After the military disaster at Pearl Harbour in 1941, Franklin Roosevelt's administration approved the lease of countless aeroplanes and ships in order to produce the movie *Air Force*, released for general viewing in 1943. The film was intended to reassure Americans that the country would be able to combat Japanese attacks on domestic territory. At the end of the film, the American airforce was shown to have triumphed (two years prematurely) over the Japanese war machine. In a similiar fashion, Walt Disney commissioned the production of *Victory Through Air Power* (1942) in order to demonstrate to American citizens the strategic significance of aircraft and their role in the American war effort against Japan and Germany. The American director Frank Capra was instrumental in this role as executive producer of the *Why We Fight* films in 1942 and 1943, which were designed to strengthen American troop morale and to explain the reasons for American involvement in the conflict with Japan. During the Second World War, Capra commanded the 834th Signal Division of the US Army and later used Nazi propaganda films to construct anti-fascist narratives and films such as *Prelude to War*. Unsurprisingly, the release of these films coincided with a rapid increase in the internment of Japanese-American citizens in the United States (Ó Tuathail 1994: 541).

In the post-war period, Hollywood continued the comfortable relationship with the Department of Defense and the Pentagon. During this period, the US military established a public relations office in Beverley Hills, Los Angeles, in order to consider producers' requests for equipment and special assistance. Two types of film project which appeared to engender considerable support from American governments concerned the Second World War and the post-war threat posed by the Soviet Union. In the war film *The Longest Day* (1961) the Defense Department gave their permission for around 700 troops to be used as film extras during Darryl Zanuck's reconstruction of the Normandy beach landings in 1944. However, due to the developing Berlin crisis in 1961, Secretary of State Robert MacNamara ordered a cut in this film contingent to 250 personnel, in the light of fears that the Soviet Union was about to invade the Western sectors of the city.

In the ideological struggle against the Soviet Union, the Eisenhower government gave permission for the Central Intelligence Agency (CIA) to participate in the production process of the first feature-length British animation film based on George Orwell's *Animal Farm* (see 'Cartoons and the antigeopolitical eye'). Under the direction of John Halas and Joy Batchelor, filming began in 1951 and was completed by 1954. The original idea for the film came from the American film producer Louis de Rochemont, who was linked to the CIA-funded Campaign for Cultural Freedom, which was designed to produce and circulate material sympathetic to the American 'way of life' based on democracy, market economics and the freedom of choice (Whitfield 1991). The involvement of the CIA in the actual production process was mainly concerned with the ending of the film. In the orginal story by George Orwell, the triumphant animal revolutionaries (mostly pigs) become the new elite and behave in a manner reminiscent of their human masters. The book ends on a pessimistic note, with the other animals (such as horses, chickens and cows) meekly accepting their new conditions without active resistance. In the film version, the animal leaders (Napoleon and his dogs) are shown to have been overthrown by the other animals, who are disaffected with their corrupt ways of governance. For the Psychological Strategy Board of the CIA, this ending of the film was crucial because it demonstrated that new forms of oppression (read the post-war Soviet Union) could be overthrown if the oppressed (read Eastern European states such as Czechoslovakia, Poland and Hungary) were organized and prepared to struggle.

The relationship between Hollywood and the US military became more problematic during the 1960s: a decade dominated by Vietnam and the US civil rights movement. The only significant film of the conflict made by US producers during the 1960s was *The Green Berets* (1968); it was directed by John Wayne, who also played one of the main characters. However, even his presence failed to allay the fears of some senior US soldiers who worried about the film's portrayal of the violent struggle in the jungles of Southeast Asia. The timing of the film was controversial too because President Johnson had already admitted the war was a 'bitch' that had caused considerable damage to American prestige and standing around the world. During the screenplay a Green Beret soldier lectures a journalist about the need to resist 'the intentional murder and torture of innocent women and children by the communists. . . . I tell you these people need us, they want us' (cited in Pilger 1998: 561). The film, however, grossed US$8 million in 1968–1969 and was considered to be a morale-boosting movie in the wake of the 1968 My Lai massacre, which led to the killing of hundreds of unarmed Vietnamese men, women and children by American troops commanded by Lieutenant William Casey.

A decade later, film-makers like Francis Coppola were denied cooperation by the American Defense Department, and the film *Apocalypse Now* (1979) was produced in the Philippines with the assistance of the Philippine armed forces. Ironically, the infamous helicopter attack scenes (accompanied by the music of Richard Wagner) were nearly cancelled, as President Ferdinand Marcos demanded the return of the helicopters because they were urgently needed for some 'real' military action in another part of the islands. During the filming process, the main actors and the director were alleged to have been profoundly disturbed by the experience of recreating the fighting scenes.

The so-called Vietnam syndrome was coined in the 1970s in response to fears from American political elites that the humiliation in Southeast Asia had caused widespread feelings of depression, guilt and a loss of moral purpose. It also led to a reluctance of further involvement in the affairs of poorly understood and distant places. After the Vietnam War, Hollywood produced numerous films in the 1980s, such as *Platoon*, *Hamburger Hill* and *Born on the 4th of July*, which sought to sympathetically portray the harrowing experiences of American soldiers in Vietnam and on their return to the United States. The popularity of these films belies the reaction of mainstream American society, which shunned returning Vietnam veterans in the 1970s (but see the film *Rambo*).

More recently the movie *Top Gun* (1986), featuring a successful US navy pilot (code name Maverick), was actively supported by the American armed

Figure 4.4 Patrolling the Persian Gulf, a US warship is pictured a few weeks before the build-up of American and British forces for Operation Desert Fox in December 1998. Since 1991 the United Nations has authorized weapons inspectors to enter Iraq and dismantle any weapons of mass destruction. Desert Fox was an Anglo-American operation intended to further damage Saddam Hussein's armed capacity. (*Photo*: Marcus Dodds)

forces as it was seen to present a positive image of the navy and its aviators. The American navy supplied a number of F-15 aircraft and the carrier *Enterprise* for the duration of filming (Figure 4.4); by the end of 1986, the film had earned US$130 milllion and eventually grossed US$350 million in world-wide cinema sales (Kellner 1995: 80). The media critic Douglas Kellner has argued that *Top Gun* was indicative of a particular period of American foreign policy, characterized by

> aggressive military intervention in the Third World, with an invasion of Grenada, the US-directed and financed Contra war against Nicaragua, the bombing of Libya, and many other secret wars and covert operations around the globe. Hollywood films nurtured this militarist mindset and thus provided cultural representations that mobilised support for such aggressive policy. . . . The 'enemy' [in *Top Gun*] flies MIGs, a Soviet plane, but is not identified as Russian, though the MIG fighter pilots have red stars on their helmets. (Kellner 1995: 75)

This interpretation of *Top Gun* is widely shared by media scholars, among them Shohat and Stam, who have argued that this film along with others such as *Rambo* epitomizes the Reaganite 1980s ethos of militarism, anti-intellectualism and social conservatism (Shohat and Stam 1994).

In contrast, *Courage Under Fire* (1996) did not enjoy the military assistance given to the producers of *Top Gun*. The request for the lease of M1 Abrahams tanks, Bradley fighting vehicles and Blackhawk helicopters was not met by the US armed forces. Media analysts considered this film noteworthy because it was the first American production to deal with the country's participation in Operation Desert Storm during 1990 and 1991, a UN-sponsored campaign that witnessed the triumph of the United States and its allies over the Iraqi forces of Saddam Hussein. However, the American military authorities responsible for cooperation with Hollywood productions rejected the call for assistance because the film dealt with the controversial issue of 'friendly fire' and attempts by senior officers to cover up the episode during the campaign. There were also demands that the main character (the Black actor Denzil Washington) did not appear drunk while wearing a military uniform. When the producers refused to alter their script to the extent demanded by the US military, they were forced to import twelve British tanks and two Cobra helicopters in order to reconstruct the battle scenes.

Films can be a very rich and varied source of research for political geographers, as well as film critics, due to the widespread distribution of particular productions and the massive audience potential. Debates about the influence and the connections between image and real-life international political behaviour continue over particular films such as *Rambo*. In terms of international affairs, there have been many examples of governments and leaders using the cinema to manipulate public opinion, often in times of crisis or war. The interactions between governments and film-producing centres have been substantial and sometimes prone to subtle interference in the final production, as the examples above have illustrated. These connections are meaningful because the US military will not lend equipment such as planes and tanks to film projects it considers unsympathetic to the armed forces. Sometimes particular representations of war and 'threats' coincide with real-life events, such as the release of the film *Black Rain* in 1990 (starring Michael Douglas, Andy Garcia and Ken Takakura) at a time when America was locked into a bitter trade war with Japan. Filmed in Japan, the screenplay concentrates on the struggles of two US detectives attempting to arrest a Japanese mafia figure among the violence and strangeness of a gangland-riddled Japanese society. The uneasy relationship between America and Japan is a central theme in the film, as even the title 'black rain' refers to the fallout from the nuclear bomb that fell on Hiroshima in 1945 (Morley and Robins 1995: 161).

More recently a new film called *The Siege* (released in November 1998) opened to considerable controversy in the United States because it featured 'Islamic terrorists' operating in Brooklyn, New York, and then being pursued by a powerful agent (played by Bruce Willis) who was appointed by the US government to crush the bombing operations. The Council of American–Islamic Relations (CAIR) and American-Arab groups have complained that the images of terror perpetuate ethnic and place-based stereotypes about Islam and the Middle East. Moreover, in the light of the ongoing negotiations between Arabs and Jews over the future of Israel and the West Bank, Palestinian commentators such as Edward Said (professor of comparative literature at Columbia University in New York) have warned that these representations of Muslims and Arabs play a part in shaping wider public attitudes towards real-life political negotiations.

A word of caution is due when considering the interpretation of films and their possible cultural and political influence (even though the analysis of audience reaction may well be possible through audience surveys and film media critiques) as there is no guarantee that the viewing public will adopt the meanings the directors and politicians have anticipated. For example, in *Top Gun* it is possible to develop a range of scenarios. The aircraft used in the attack scenes appear to be MIG jets but were actually American Northrop Freedom fighters, and the enemy's red stars need not necessarily be Soviet but could refer to various Arab and/or Chinese armed forces. The red star is highly symbolic because it is identified as referring to communist countries and socialist ideologies. Yet *Top Gun* could be interpreted as a film supportive of the aggressive military action of the United States in the Middle East during the 1980s, culminating in the deployment of troops to the Lebanon. Alternatively, many viewers of *Top Gun* may not have made those connections at all and simply thought it was an action-packed film about planes, pilots and the personal relationship between Maverick and his flight instructor, Charlie (played by Kelly McGillis).

Film and cinema offer exciting possibilities for considering the prevalent geopolitical representations of world politics and places. The popular movie cultures generated by Hollywood clearly have a tremendous impact in terms of audience figures and revenue generation. However, the interpretation of film (and, for example, the analysis of 'threat' construction) needs to be treated with caution as there is no automatic or causal link between the film and audience reaction. While the geopolitical significance of film can be overrated, recent controversies surrounding *The Siege* and Oliver Stone's proposal to make a film about the TWA airliner that crashed off the North American coastline in July 1996 – Stone's proposed script ventured that the US Navy had shot down the plane by mistake; ABC, the sponsoring network, dropped the project – nevertheless illustrate the contested and controversial nature of film.

'Something must be done!'

The 1980s and 1990s witnessed the growth of globally orientated events such as the 1992 Earth Summit and the increasingly high-level coverage of war and humanitarian disasters such as the 1984 Ethiopian famine and, more recently, the terrible consequences of Hurricane Mitch on Central America in October/November 1998. The power of the television image was recognized by the Canadian theorist Marshall McLuhan in the 1950s; yet despite the exponential growth of television ownership and media networks in the Western world and beyond, the theme of his work remains as poignant as ever. The apparently simple observation that the medium, i.e. the mode of delivery, is the message remains richly suggestive (Crang 1998: 93–94). Television provides a potentially revolutionary medium for information exchange because of its ability to transcend spatial and social boundaries around the world.

The relationship between television coverage and international politics has recently attracted much critical attention in media studies, international politics and geopolitics. Major media networks such as Cable News Network, Time-Warner, News Corporation and Sony have emerged during the 1980s and 1990s, and demonstrated a remarkable capacity for shaping international agendas. The advent of portable video cameras and satellite dishes and the increase in freelance journalism has further contributed to this phenomenal growth, and it is estimated that around two hundred television channels will be licensed to operate in the United States by the end of the present century (Morley and Robins 1995: 13). The speed and range of contemporary television coverage is such that media observers have often referred to the rise of the so-called CNN factor. This is a reference to the 24-hour Cable News Network channel based in Atlanta, Georgia, which has provided a worldwide English-speaking news network since 1980. In conjunction with other media empires such as Time-Warner and News corporation, CNN occupies a powerful position in terms of audience figures and news production within the global broadcasting world (Barker 1998).

Globalization theorists often refer to the rise of real-time television networks as further evidence of time–space compression, as the experiences of distant places are brought into the living rooms of principally Western citizens. The BBC's coverage of starving children during the 1984 Ethiopian famine was believed to have been the impetus for television viewers to contribute to the relief appeals. The release of the song 'Do they know it's Christmas?', recorded by a group of artists under the name Band Aid, became one of the biggest-selling singles in 1984, in direct response to their commitment that all proceeds from the sale of the record would be donated to the famine relief in Ethiopia.

Television coverage of special events such as famine and war can obscure geopolitical issues. In the case of Ethiopia, the news of a massive famine coincided with the intensification of the Cold War between the United States and the Soviet Union in various areas of the world, including sub-Saharan Africa. Viewers were presented with ever more distressing pictures of starving children yet relatively little coverage was devoted to the wider issues concerning the origins of the famine, such as the political and economic context of the country. This humanitarian disaster was not entirely due to the failure of the 'rains', but was undoubtedly exacerbated by the ongoing civil war between rival factions (supported by either the Soviet Union or the United States) and the fact that the fertile lands of Ethiopia were being used to grow cotton to enable the ruling government to secure export earnings in order to purchase further weapons (O'Loughlin 1989). Apparently, after the American NBC channel covered the famine story for American viewers, the aid given to Ethiopia by the Reagan government increased from US$20 million to US$100 million during the course of 1984 (Harrison and Palmer 1986).

The experience of the 1984 Ethiopian famine and other such notable events raised a number of controversial issues with regards to the role of television and its influence on political decision making and public opinion. The first problematic was advanced by the French philosopher Jean Baudrillard, who published a series of provocative articles in the French newspaper *Liberation* from January 1991 onwards. In the case of one particularly famous article, 'The Gulf War did not happen', Baudrillard argued that the massive increase in television images over the last thirty years has changed our relationship to the wider world. More specifically, that television coverage of the 1991 Persian Gulf War was increasingly comparable to a movie or a series of images rather than the culmination of the coverage of real-life events, such as the bombing of Iraqi troop positions, power stations in Baghdad and other places in Iraq. In an often misunderstood argument, Baudrillard is not suggesting that Operation Desert Storm did not happen, but rather that the video images of the Gulf campaign became more significant than the real-life events. Viewers were increasingly encouraged to watch images of the bombardment and compare video stills rather than to actively contemplate the horrific consequences on human life of so-called smart bombs and cruise missiles.

Television coverage of international politics, although capable of exposing injustice and mobilizing public opposition to brutal state violence, does not perform such an emancipatory role. Television pictures of the Chinese student protests in Tiananmen Square and the resultant massacre by the Chinese armed forces in June 1989 did not provoke other governments to take preventive or retaliatory action. In that sense, Bernard Cohen's comments about the value of media reporting in Somalia during the 1992 UN operation overstate the case:

> Television has demonstrated its power to move governments. By focusing daily on the starving children in Somalia, a pictorial story tailor-made for television, TV mobilised the conscience of the nation's public institutions, compelling the government [i.e. the US administration] into a policy of intervention for humanitarian reasons. (Cited in Mermin 1997: 385)

The more recent research of the role television coverage exerts in shaping American foreign policy has concluded that 'Somalia' (the scene of a UN operation in 1992) had already become a humanitarian issue within the foreign policy community in Washington before the camera crews arrived in Africa. As in the Bosnian wars of the 1990s, television provides popular geopolitics with interesting examples of how foreign policy and humanitarian missions may be shaped by its coverage, and why films, audience reaction and political significance must be carefully examined rather than simply assumed.

Reader's Digest and the Cold War

The capacity of popular magazines to represent specific geopolitical spaces such as the 'Soviet Union' or the 'Far East' via the construction of particular subjectivities, encourages the reader to identify with a host of ideas ranging from consumer issues to international political agendas. Not all readers will interpret the stories in the same fashion; indeed it is often the case that one

Figure 4.5 A 1993 cover for UK *Reader's Digest*: note the monthly circulation of 28 million copies worldwide. (Reproduced by kind permission)

particularly dominant interpretation or reading will emerge as hegemonic, depending on the subject matter. In the case of the Cold War period, American magazines often employed a mixture of discourses, including individualism, morality, manifest destiny and American exceptionalism, in order to promote geopolitical visions of 'America' to other geopolitical relations and representations.

Popular magazines and journals such as the *Reader's Digest*, *Life* and the *National Geographic* have been long-standing features of Euro-American public life in the present century. This assessment was based on a variety of criteria ranging from circulation figures to the political and cultural significance of the contributing authors. The *Reader's Digest*, for instance, enjoys a circulation figure of 16 million in the United States and is widely translated from English into a variety of other European and non-European languages (Figure 4.5). The founder of the magazine, DeWitt Wallace, had a particular geopolitical vision, which was deeply sceptical of the Soviet Union, trade unions, totalitarian governance and communist politics. The Italian version of the *Reader's Digest* was launched in 1948, when it was feared that the Italian Communist Party would seize power following democratic elections (Sharp 1996: 567). One of its remarkable features was the ideological transformation of the Soviet Union from wartime ally to post-war adversary.

This shaping of American attitudes towards the Soviet Union in the Cold War period was demonstrated in a series of papers by the British geographer Joanne Sharp (1993, 1996). Since the 1920s, the unsettling and potentially threatening character of the Soviet Union generated a steady stream of articles on the totalitarian nature of Soviet political life. Sharp argues that the US version of the magazine participated in maintaining a particularly dominant representation of the Soviet Union throughout the Cold War period and actively encouraged the reader to identify with American interests at the expense of the 'Soviets' or 'Reds', who were portrayed as a threat to the 'American way of life'. These geopolitical strategies have also been noted by others:

> Now readers were bombarded with accounts of the total difference with and opposing character of the Soviet Union to the 'actual existing' United States, not simply as the actual opposite of the other. Such representations had powerful effects in mobilising public opinion into the Cold War consensus that progressively engulfed American politics from 1947 until the Vietnam War of the late 1960s. (Agnew 1998: 110)

On the domestic front, the *Reader's Digest* portrayed a particularly narrow vision of American political identity based on the belief that the American economy and society must be supported, among other things, by mass consumption and production, an enlarged military presence in the wider world and complete hostility to any form of left-wing or labour militancy. As the American political scene of the 1950s and 1960s demonstrated, free and competitive democratic politics was not possible in a climate of profound paranoia towards socialism and communism. As Ó Tuathail and Agnew concluded, 'The simple story of a great struggle between a democratic "West" against a formidable and expansionist "East" became the most dominant and durable geopolitical script of the [Cold War] period' (1992: 190), which was reproduced in countless forms of media and cultural expressions ranging from films and television broadcasts to popular magazines and journals.

The *Reader's Digest* participated in the construction and legitimation of a particular Cold War geopolitical imagination which witnessed the division of the world into competing blocs. Stories about communism and the 'Soviet way of life' were counterposed with representations of 'America' that tended to emphasize not only national territory but also a transcendental morality, an idea which meant the United States had a manifest duty to protect democracy and capitalism for the wider world. In a trend that echoed earlier episodes of American exceptionalism, 'America' became transformed into a symbolic space (a beacon on the hill) which had a moral responsibilty to protect and promote its own ideological and material vision for the post-war world (Agnew 1998).

Ironically, the *Reader's Digest* was extremely sceptical of Gorbachev's reformist government in the Soviet Union during the mid-1980s (Sharp

1993). In the period after the Cold War, the magazine continued to warn its readers that the Soviet Union might not be capable of change in terms of totalitarianism and the apparent rejection of market economies. This view might yet be vindicated given the current troubles in Russia over a collapsing currency and the re-emergence of the Communist Party.

However, it also raised the possibility that America was now threatened with other dangers, ranging from domestic terrorism (such as the 1995 Oklahoma City bombing by far right fanatics) and Islamic fundamentalism in places such as Iran and Libya, to the economic threat of Japan. In contrast to the Cold War, the geographies of danger appear more varied and complex. During the 1990s *Reader's Digest* warned its readers that dangers and threats lay within and outside the boundaries of the United States.

Cartoons and the antigeopolitical eye

Political cartoons have a long and varied political and artistic history ranging from the English tradition of political satire (including figures such as Gilray, Strube, Low and Rowlandson) to the anarchical farce of contemporary artists such as Steve Bell and Matt Wuerker (Ó Tuathail, Dalby and Routledge 1998: 3). In recent years, political geographers have increasingly appreciated that cartoons can be considered in conjunction with other popular sources such as magazines, newspapers and television. This has also been matched with a growing interest in cultural geography for visual sources combined with a commitment to analysing cartoons and images as 'texts' capable of multiple interpretations (Crang 1998).

At first glance, the significance of humour and cartoons to international affairs may appear either remote or simply rather silly. It is quite common for students of world politics to produce weighty tomes on the condition of the interstate system and foreign policy without any illustrations or images. However, a number of authors have argued that political cartoons and images can be deployed as 'geopolitical texts', which illuminate or even subvert particular political practices such as foreign policy decision-making. Political images can also be deployed to illustrate prevalent cultural anxieties about 'threatening neighbours' and 'dangers'. Critical geopolitical writers have also argued that cartoons can question and even transgress dominant relations of knowledge, truth and power. Cartoonists such as Steve Bell deploy an antigeopolitical eye as they contest dominant representations of political affairs such as the 1982 Falklands War and the Bosnian crisis of 1992–1995 (Dodds 1996, 1998).

The term 'antigeopolitical eye' was used by Gearoid Ó Tuathail to describe the critical writings and images of journalists such as Maggie O'Kane and cartoonists such as Steve Bell. Some of Bell's images are unsettling precisely because he uses satire, humour and shocking representations to question specific events such as the 1982 Falklands War or the Bosnian crisis. The best images and articles are acts of transgression in that they question the dominant relations of power, knowledge and truth. Bell's images of Bosnia, for example, refused to accept the British government's assertion in 1993–1994 that intervention on behalf

Figure 4.6 When Allied shipping was sunk in the North Atlantic during the Second World War, Philip Zec's *Daily Mirror* cartoon (1942) proved one of the most controversial to be published during the entire conflict.

of vulnerable civilians was politically unsafe and strategically unwise. His shocking portrayals of death and destruction helped to restore 'Bosnia' to our universe of obligation. As visual critiques, cartoons help to deconstruct the political agendas of political elites, national security bureaucracies and military officers.

In the midst of the Second World War, the *Daily Mirror*'s cartoonist Philip Zec published a famous cartoon of a ship-wrecked sailor clinging desperately to a piece of wood while floating in the middle of an ocean (Figure 4.6). The caption of the image read 'The price of petrol has risen by a penny (Official)'. The cartoon coincided with a particularly traumatic moment in British politics when in 1942 the *Daily Mirror* attacked the retention of ministers in Churchill's coalition government, ministers who had earlier proposed to 'appease Hitler' in 1938. Prime Minister Winston Churchill then ordered a financial investigation of the *Daily Mirror*'s affairs in the light of criticisms of his ministers, including Austin Chamberlain. The apparently unpatriotic behaviour of the newspaper was further condemned when Zec's cartoon was interpreted as being highly critical of the government's ability to deal with the German naval attacks on British shipping in the North Atlantic. The home secretary, Herbert Morrison, came to Parliament and complained that, 'The cartoon in question is a particularly evil example of the policy and methods of a newspaper with a reckless indifference to the national interest' (cited in

Figure 4.7 A 1950s cartoon produced under the auspices of the Foreign Office International Research Department. (Reproduced with permission of the Public Records Office, Kew, London)

Pilger 1998: 387). Morrison believed the cartoon suggested that the lives of merchant seamen were being sacrificed in favour of the oil companies, who would charge more for their products at a time of national shortage. Zec maintained that he saw the image as highlighting the need to save petrol because importing goods and products cost lives. However, the *Daily Mirror's* reputation for frank criticism meant that this cartoon became one of the most contested British images of the Second World War.

George Orwell's *Animal Farm*, published in 1945, was a thinly disguised critique of the Soviet Union and the corruptive practices of totalitarian governance. Given the timing of its publication, the British Foreign Office recognized the potential of *Animal Farm* as a source of propaganda against the Soviet Union. Officials in the International Research Department of the Foreign Office also commissioned Orwell to produce a cartoon strip of the book in order that British embassies around the world could supply them to local newspapers. It was noted in one Foreign Office memo dated from 1951 that, 'With a skillful story-teller one should have thought that it could be made into a very effective piece of propaganda down to *village audience level*' (my emphasis and cited in Norton-Taylor 1998: 7); see Figure 4.7.

The phrase 'village audience level' becomes all the more evident if one considers what cartoons can achieve in a performative sense. Cartoons are a specific media form using humour and satire to convey messages about the social and political world (Dines 1995: 237). The cartoon illustration of Orwell's *Animal Farm* shows a disaffected pig called Major dreaming of revolution (Animals arise!) because of the apparent indifference of Farmer Jones to the welfare of his animals on the farm. British officials were convinced that Orwell's images were useful because they would appeal to the semi-literate 'village populations' of the strategically important countries such as India, Mexico and Brazil. Notwithstanding the often patronizing assumptions made about these audiences, the warning of the cartoons and the book appeared to be profound: Western capitalist societies could not be complacent against the threat of Soviet totalitarianism, and totalitarianism was riddled with dangerous contradictions.

The recently declassified secret papers of the British Foreign Office have revealed that George Orwell was not only responsible for naming other artists and writers as 'crypto-communists' but also for encouraging the then British government to engage fellow writers in a visual propaganda campaign against the Soviet Union. During the 1950s, a series of artists and writers were commissioned to produce cartoons and illustrations which sought to represent the Soviet Union as a dangerous and threatening place to Britain and her allies. Officials within the International Research Department of the Foreign Office also constructed other characters in order to illustrate the apparent dangers of communism and totalitarianism. One such example was the invention of a character called Guy Greenhorn who was depicted as a citizen of 'Democrita' and a 'likeable, intelligent but gullible young fellow and an admirer and sympathiser of communism'. He believed in the principles of 'Stalinovia', but was vulnerable to the influence of the evil-looking Dr Renegado. Greenhorn meets a young woman, but eventually they become disenchanted with married life in Stalinovia. The tale concluded with the young Greenhorn and his wife returning to Democrita because they could no longer tolerate the restrictions of Stalinovia.

Perhaps this appears rather bizarre and even comical to readers in the late 1990s, yet remember how British and American political life in the 1950s was dominated not only by the hope for post-war economic recovery but also by a fear of communism and the Soviet Union. The illustrated story of Greenhorn, which was released by the Churchill government in Britain, coincided with a series of high-profile defections to the Soviet Union by Cambridge-educated Foreign Office officials such as Guy Burgess and Kim Philby. In a state of near paranoia over intelligence leaks and communist penetration, cartoons and illustrated stories were seen as a source of geopolitical material in the struggle against a Cold War adversary.

The spatial symbolics of humour can often challenge and even subvert dominant boundaries of national sovereignty and the nationalist scripting of place. During the 1982 Falklands War, for example, many editors felt obliged to adopt a position broadly supportive of Mrs Thatcher's decision to launch a task force to recover the Falkland Islands in April 1982 (Figure 4.8). The then government had invested considerable political and cultural capital in persuading editors and public opinion that although the Islands were located 8,000 miles away from Britain, they were in fact populated by loyal and patriotic British citizens. As Peter Jenkins of the *Guardian* noted in April 1982, 'By what weird calculus was it reckoned that the fate of all the free peoples might hinge upon the fate of 1,800 islanders and their 600,000 sheep?' (cited in Dodds 1996: 572). Alternatively, cartoonists such as Bell used cartoon images to depict the Falklands as a group of windswept rocks overwhelmingly populated by sheep and penguins. The presence of sheep could therefore be a source of familiarity for readers or a means to ridicule the British response to the loss of these Islands at a time of high unemployment and social dislocation in the United Kingdom.

Figure 4.8 Steve Bell's cartoon strip 'If . . .' appeared in the *Guardian* during the 1982 Falklands/Malvinas War. (Reproduced with the permission of Steve Bell)

Geopolitics and music

As a form of human expression, music has been a powerful vehicle not only for the articulation of dissent and resistance (e.g. popular songs such as 'Free Nelson Mandela' and 'Sunday Bloody Sunday') but also as a tool used by regimes and governments to gather support for particular forms of nationalism, sometimes extremely violent and xenophobic (e.g. Nazi Germany in the 1930s used the stirring music of the German composer Richard Wagner to engender a sense of nationalism and patriotic loyalty). Popular musicians in America such as Robert Zimmerman (Bob Dylan) were at the forefront of anti-Vietnam protests in the 1960s. More recently, music and musicians have attracted considerable controversy, as witnessed during the 1991 Gulf War when British DJs were instructed not to play songs such as Abba's 'Waterloo' for fear of causing offence to the Allied troops stationed in Kuwait and/or their families. The lyrics of the song were considered not only excessively militaristic but also potentially insensitive to French troops serving in the Gulf, because 'Waterloo' refers to an Anglo-German victory over the French in 1815. At the same time, musicians in Algeria and Afghanistan were murdered because the extremist authorities believed that music was a form of Western pollution which bred dissent and disrespect for Islam and the Islamic way of life.

Anglo-American geographers have recently begun to explore how music contributes to specific constructions of place and cultural identity (S. Smith 1994; Leyshon, Matless and Revill 1995). In particular, this tranche of geographical research has emphasized how music (from classical to rave) has been neglected in the social sciences and humanities in favour of more visual sources such as film, cartoons, television, paintings and landscapes (S. Smith 1994: 235). In the last five years, however, geographers have produced some interesting research which seeks to locate music in its geographical, political and cultural contexts. In nineteenth-century Britain, for example, the brass band movement forged and sustained community music making in industrial areas which depended on extractive industries such as coal and oil, iron and steel. In the aftermath of industrial decline, the brass band movement helped to consolidate communal loyalties (and a collective sense of purpose) when mass unemployment and social dislocation occurred in the late twentieth century (S. Smith 1994).

Human rights organizations consider music to be one of the most censored forms of art. During the Cold War, for instance, dissident bands in communist Eastern Europe (such as Plastic People of the Universe in Czechoslovakia) released their music via underground recording labels in order to protest about all forms of censorship and human rights violations. Popular music was considered especially subversive because of its appeal to young people. The present Czech president, dissident writer Vaclav Havel, revealed in 1989 how he had been inspired by the music of Lou Reed. In South Africa, the American musician Paul Simon joined forces with the Black band Ladysmith Black Mambazo to protest through music at the continued injustices and inequalities of apartheid. During the repressive era of General Pinochet's Chile (1973–1989), the security forces killed the folk singer Victor Jara in a Santiago football stadium in 1975 because he was considered a left-wing subversive.

Thus far, geopolitical writers have not fully explored the geographical soundscapes and political worlds created by music. A case for consideration would be the music of the Irish band U2, created and sustained by a particular representation of British violence in Northern Ireland. Their best-selling album *Under a Blood Red Sky* (1981) opens with the now immortal line 'This is not a rebel song, this is Sunday Bloody Sunday'. It is ironic, because the song (Sunday Bloody Sunday) refers to the shooting of thirteen unarmed nationalist (Catholic) civilians by British paratroopers in January 1972 (called Bloody Sunday by the media) in Northern Ireland. For many nationalists this massacre further cemented the view that Northern Ireland was a province controlled by a Protestant/Unionist majority using the armed forces and police service to oppress and even murder a religious minority. For Irish people (from the north and south of the island of Ireland), rebel songs (many originating from the bloody civil wars of the 1920s) have served to express dissent and resistance against the continuing British colonization and occupation of Northern Ireland.

Alternatively, punk bands in Germany generated a following of sorts among young (White) Germans attracted to lyrics blaming unemployment and poverty on immigrants and foreigners. While it would be wrong to imply this music directly contributed to outbreaks of violence against Turkish families (as has happened in Germany to German-Turkish residents), there is no doubt that music can evoke powerful feelings and provoke movements and acts of political resistance. Music can help to create particular political and symbolic geographies of resistance, censorship and expressions but it can also be a matter of life and death in countries such as Algeria, Myanmar and China. The death of the singer Boudjema Bechiri in Algeria in 1996 is one of the saddest examples of a musician being murdered by a brutual government seeking to suppress any form of dissent or resistance in the midst of Algeria's bloody and ongoing civil war.

Conclusion

This chapter has illustrated how the cultural media can be used to examine various forms of communication and imagery to represent the social and political world. The usefulness of films can also be gauged through the varied geographies of dissemination and distribution, which ensure movies such as *Top Gun* are seen not only in the cinema but also via commerical television and video rental. The role of the televisual media in representing and portraying international events ranging from war, humanitarian disasters and international conferences to televisual diplomacy has become commonly accepted as pundits and observers recognize the power of television pictures and soundbites within world politics. In a similiar vein, magazines and cartoons also contribute to the communication of ideas and intepretations of world affairs along with specific places and peoples.

Geopolitics embraces various media and institutions as well as the role of monuments, public education and literary novels in the construction of geopolitical imaginations. Geographical instruction, for instance, seeks to promote forms of patriotic education in countries such as Argentina and Chile, where the territorial boundaries of the state have been perceived to be under threat by neighbouring states. State schools in combination with the media and military cartographic institutions endeavour to distribute representations of national space which either exaggerate territorial losses and/or attach special meaning to certain places. In Argentina, for example, the Falkland Islands are described as the Islas Malvinas, and countless posters, maps and billboards proclaim that the islands belong to the Argentine people not the British state. The loss of the Malvinas has therefore assumed a special symbol within the collective Argentine geographical imagination.

This chapter has not considered the information superhighway as a popular geopolitical source, but it is clear that the Net has the capacity to transcend the regulations of states by virtue of being apparently beyond national control. In certain circumstances this can enhance the capacity of NGOs, such as Amnesty International, to circumvent particularly repressive regimes by publishing details of human rights abuses without direct interference. Alternatively, the unregulated flow of child pornography pinpoints the dangers associated with the Net. At present, access to new technologies is profoundly limited as most of the users and providers of information reside in North America, Western Europe and East Asia. None the less, as geographers and social scientists more generally recognize, the Net does provide new opportunities for redefining political space, political communities and virtual landscapes.

Further reading

There is a massive amount of literature available on the various forms of media discussed. For film and television in general, see P. Harrison and R. Palmer's *News Out of Africa* (Hilary Shipman, 1986) and C. Barker's *Global Television* (Routledge, 1998). On representation and the cinema, see E. Shohat and R. Stam's *Unthinking Eurocentrism* (Routledge, 1994) D. Morley and K. Robins' *Spaces of Identity* (Routledge, 1995) and K. Robins' *Into the Image* (Routledge, 1996). On the importance of cartoons, see G. Dines, 'Towards a sociological analysis of cartoons', *Humor* **8**: 237–55. On the analysis of music, see A. Leyshon, D. Matless and G. Revill, 'The place of music', *Transactions of the Institute of British Geographers* **20**: 423–33. On the geopolitical significance of popular sources see *The Geopolitics Reader* edited by G. Ó Tuathail, S. Dalby and P. Routledge (Routledge, 1998) and M. Crang's *Cultural Geography* (Routledge, 1998).

Chapter 5

The globalization of danger: nuclear weapons and nuclear testing

The threat posed by nuclear weapons is enormous, as even isolated experimental nuclear explosions have the potential to destroy parts of the earth's ecosystems. The deadly power of nuclear bombs gave rise to a host of anti-nuclear movements in North America and Western Europe. They argued that Cold War ideologies of nuclear deterrence contributed to further insecurity in the world, with the creation of a nuclear culture that dehumanized opponents, exaggerated threats to national security and downplayed the consequences for human and environmental life (Lifton and Falk 1982; Beck 1992). The effects of test programmes on people living in the South Pacific, the centre of Australia and the allegedly remote parts of the United States and the former Soviet Union have only been formally documented in the last ten to fifteen years. For many people affected by nuclear testing, nuclear war was not unthinkable.

In our globalizing world, the destructive powers of nuclear weapons appear to remain a graphic symbol of a fragile planet potentially threatened by mass destruction. Within a geopolitical framework, the spread of nuclear weapons and the geographies of nuclear testing provide evidence of its manifestation on global politics:

- For realists, nuclear proliferation demonstrates how powerful states have sought to develop nuclear diplomacy in order to protect their national security interests.
- For the liberal institutionalists, the capacity of the international community and organizations such as the United Nations to deal with nuclear weapons has been stretched. Overall, however, the United Nations has managed to create a negotiating atmosphere conducive to a measure of control and disarmament (Keohane and Nye 1989).
- For feminist observers, both realist and liberal institutionalist analyses of nuclear weapons and proliferation neglect how the experiences and consequences of these issues vary between men and women and from place to place. It has also been noted how the geographies of nuclear testing were felt most strongly in the territories of tribal peoples and ethnic minorities (Enloe 1989, 1993).
- Critical geopolitical writers have highlighted how nuclear weapons raised important questions about the nature of global political life and the unequal geographies of nuclear threats and testing/testers (Dalby 1990).

In the post-Cold War era new hopes emerged that nuclear weapons could be abolished if the superpowers agreed to major cuts in their stockpiles and to the removal of weapons systems from Western and Central Europe. Sceptics argue that nuclear proliferation still remains a serious problem for global politics, as witnessed by the ongoing attempts of Israel, North Korea, India and Pakistan to develop a nuclear capability and the discovery of Iraq's clandestine nuclear weapons programme in the aftermath of Operation Desert Storm (Gardner 1994; Mazarr 1995). On the one hand, realists claim that nuclear weapons cannot be removed from the international political scene because they cannot be 'uninvented'; but on the other hand, a number of critics insist that this sort of pessimism could be overcome by developing rigorous global disarmament regimes and by generating a global consensus to denounce nuclear and other weapons of mass destruction.

This chapter is organized into four sections. First, the significance of nuclear weapons is discussed in global politics and the role played by the United Nations in maintaining international awareness of nuclear weapons. Second, the varied attempts to control horizontal and vertical nuclear proliferation in the post-war period are reviewed and the regional initiatives designed to declare nuclear-free zones considered. Finally, the opportunities and dangers inherent in nuclear weapons control in the 1990s are explored. It is argued that while the Comprehensive Nuclear Test Ban Treaty has offered new hope for global nuclear disarmament, it has not resolved some fundamental tensions between nuclear and non-nuclear weapon states.

Global politics, nuclear weapons and the nuclear weapons cycle

Since 1945 the international community has faced the possibility of global nuclear war. Michael Frayn's recent play *Copenhagen* (which opened at London's National Theatre in June 1998) considered the lives of two men intimately involved in the development of the nuclear bomb. In German-occupied Denmark during 1941, two nuclear scientists Werner Heisenberg (involved with the Nazi's attempts to construct a bomb) and Niels Bohr (one of the inventors of quantum physics) meet to discuss among other things the scientific and moral ethics surrounding the bomb. It is also a play about uncertainty (Heisenberg later gave his name to a fundamental principle of uncertainty within physics), as the audience is left wondering whether Heisenberg came to see Bohr in Copenhagen in order to discover the extent of Allied knowledge about the bomb. With the men discussing the construction and implications of the bomb, Margarethe Bohr is privy to the conversation and acts as an 'interpreter' for the audience by insisting that the abstract discussions on particle movements are humanized by giving due consideration to the consequences of their actions.

Ironically, the major figures behind the first nuclear explosion in New Mexico in July 1945 were German-speaking scientists who had fled their native European countries during the Second World War, or who were recruited by the Americans at the end of the conflict. Despite the claims to benefit civilization, the United States had ensured that the twentieth century would be immortalized as a dangerous and nuclear age. The subsequent American bombing of Nagasaki and Hiroshima in August 1945 demonstrated the deadly potential of nuclear explosions. It is estimated that, in terms of immediate deaths, 70,000 died in Hiroshima and 35,000 perished in Nagasaki. The overall death toll is unknown because many of the survivors later died from ill-health relating to nuclear radiation. In 1946 the US government recruited German military scientists to assist in the development of nuclear technology, rocket and space technology. Within three years of the American nuclear bombing, the then recently created United Nations Commission for Conventional Armaments decreed a new category: weapons of mass destruction. By 1964 five countries had tested either fission or fusion nuclear weapons: the United States, the Soviet Union, the United Kingdom, China and France. The nuclear weapons cycle (NWC) refers to the production, testing and deployment of nuclear weapons. While it is common to focus on the production and deployment of such arms systems, the geographies of testing and their social and cultural implications must not be neglected by this analysis of world politics. Nuclear weapons testing occurred in many parts of the world, often in places perceived to be marginal in the minds and actions of national elites. Indeed, for the first decade of the so-called post-war period, many heads of state viewed the invention of nuclear energy as a positive sign: a source of endless and cheap energy and a symbol of the modern technological state. For example, President Truman's 'atoms for peace' speech in 1953 was similar in tone to his earlier speech in 1949 extolling the virtues of modern industrial-based development (Chapter 3).

Global and national institutions have struggled to cope with the legacy posed by nuclear weapons through either arms control or disarmament. Arms control refers to a process of gradually limiting and/or restraining production, testing and deployment of weapons, whereas disarmament is a more radical proposition concerned with the renouncement of weapon systems. In realist thought, nuclear weapons and nuclear culture have been endowed with qualities which emphasized the enhanced capabilities afforded to a state within the anarchical international arena. In that sense, the acquisition of nuclear weapons technology by the United Kingdom in the late 1940s must be considered indicative of a desire by the Churchill government to remain a major power on the world stage at the end of the Second World War. With the emergence of the two superpowers, many British public figures were arguing that nuclear weapons would improve Britain's standing in the world. By the mid-1960s, all five permanent members of the Security Council (China, France, the Soviet Union, the United Kingdom and the United States) had obtained the status of nuclear weapon state (NWS).

Realists argue that the possession of nuclear weapons offers certain advantages in terms of bargaining power and international profile. However, there is also a strong sense that the pursuit of national interest and national survival would not be facilitated by an aggressive deployment of nuclear weapons. In the words of some writers, a form of 'nuclear taboo' has existed during the last fifty years, in which states have sought to engage in confidence-building measures and arms control processes in order to lessen the dangers of a potential nuclear war. It was argued, therefore, that nuclear weapons provided a form of stability for the international system during the Cold War period.

In conjunction with liberals, realists also believe that the arms control process sponsored by institutions such as the United Nations and specialist international agencies can help to mediate the dangers posed by nuclear weapons. The UN charter commits states to pursuing peace and common security under article 11, and the United Nations has sought to advocate solutions to specific problems and to sponsor resolutions on nuclear testing, deployment and the peaceful use of nuclear energy. Both realists and liberal institutionalists endorse the significance of confidence-building as well as international diplomacy and designated nuclear conferences to generate a consensus on such issues.

For the globalization theorist, nuclear weapons provide interesting material on the spread of technology (horizontal proliferation) and danger across the planet. Having rejected the overwhelming focus on political power and the state, globalization and feminist theorists on world politics emphasized the complexity surrounding nuclear issues. Apart from social and political life, both the state and international relations have been profoundly affected by the intensity of the nuclear weapons cycle. In the post-war period, the nuclear weapon states began a cycle of extensive nuclear testing in either the Third World (or remote parts of China and the Soviet Union) or in places populated by indigenous peoples. In the period between 1945 and 1991, over 1,900 nuclear explosions were carried out on the earth's surface. The leading testers were the United States (936), the Soviet Union (715), France (192), the United Kingdom (44) and China (36). The primary sites of these tests were Kazakhstan (467), French Polynesia (167), the Marshall Islands (66), Xinjiang Province of China (36) and other sites such as indigenous peoples' reservations in Nevada and Australia. The effects of nuclear testing are still being felt in the communities of many of these islands, remote provinces and interior reserves (Kato 1993).

The impact on the communities of nuclear testing have been partially documented (Figure 5.1). Joni Seager noted, 'Most of the nuclear weapon systems stationed in Europe, and the US have been tested on "indigenous people's" lands in the Pacific, without their consent and often without warning' (Seager 1993: 61). In the 1950s, for example, Britain carried out nuclear testing in the centre of Australia and on Christmas Island in the Pacific Ocean. With the assistance of the Australian government, 800,000 square kilometres of Aboriginal land in Northern Territory was given over to nuclear testing without any form of compensation for those who were effectively dispossessed. In other parts of Australia, such as Pine Gap and Nurrungar, the Americans created top-secret surveillance bases, which were off-limits to all Australian citizens. Most of the nuclear test sites remain highly contaminated, and the British and Australian governments are reluctant to address the substantial issues of cleaning up the sites and compensating the victims.

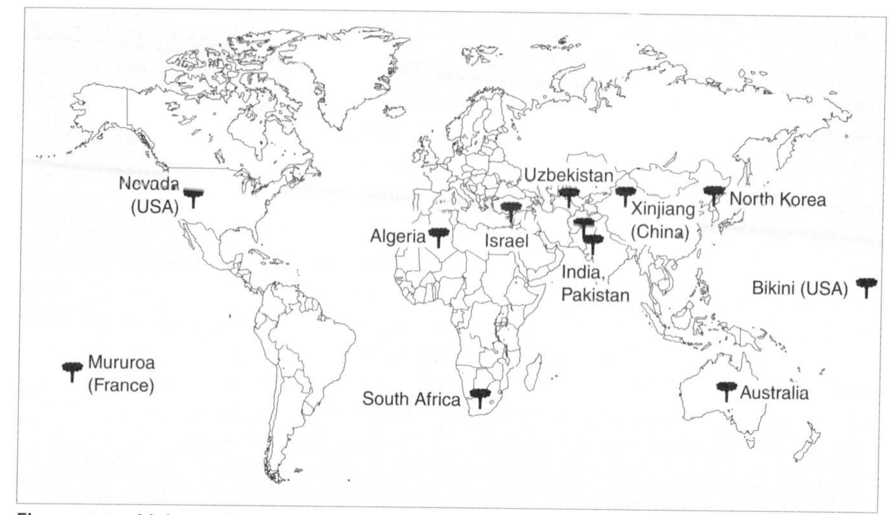

Figure 5.1 Major nuclear test sites around the world.

By a further twist of irony (see below), the decision to declare Antarctica a zone of peace and cooperation in 1959 was only made possible because the three major superpowers had secured alternative sites such as Nevada, Siberia, the Pacific Ocean and the Australian continent for test purposes. Five years earlier, in 1954, the American armed forces had bombed the Bikini Islands using 17-megaton hydrogen bombs, in order to assess the likely impact of a bomb one thousand times more powerful than those dropped on Japan in 1945. The islands have now been declared uninhabitable for thirty thousand years (Seager 1993: 64) and the original residents moved around various island groups without compensation.

Most standard realist texts on nuclear strategy and weapons have little time or scope for the local consequences of nuclear testing. The combined effects of British, French and American testing in the South Pacific would be relegated to a footnote in the accounts of weapon systems and their effect on the international scene. The key dimensions of international life remain military force, war, the state and the anarchical international arena. Feminist and antinuclear observers would argue that women and ethnic minority groups have often borne the brunt of this vision of political life (Pettman 1996). The litany of health problems in these areas is truly shocking, given the high incidence of cancers, deformed babies and premature death. Governments have often been reluctant to accept responsibility for their nuclear testing activities. At present, the Australian, the UK and the US governments are negotiating compensation packages with indigenous peoples affected by nuclear testing. While liberals have tended to emphasize the capacity and willingness of the international community to engage in negotiation and confidence-building measures, in an attempt to prevent nuclear weapons from becoming an accepted and widespread feature of global political life, critics point to nuclear testing stories to illustrate the capacity of these weapon systems to tie states and their peoples into an unequal network of social relations which span the globe.

Post-1945 nuclear proliferation control

Since 1945 both nuclear technology and ballistic missile systems have spread from a select number of nations to encompass and/or affect all the major geographical regions of the world. In a spirit of premature optimism, the United Nations passed a resolution which established the UN Atomic Energy Commission (AEC) in January 1946 and the Commission for Conventional Arms in 1947 (Whitaker 1997: 58). The aim of the Atomic Energy Commission was to promote the peaceful usage of nuclear energy while seeking to eliminate all nuclear weapons. In part, the inspiration for the AEC came from the US government, already anxious in 1946 to restrain nuclear proliferation not least because it was common knowledge that the Soviet Union was close to exploding their first nuclear bomb. After a ten-year hiatus, largely caused by worsening relations between the superpowers, the International Atomic Energy Agency (IAEA) was created in 1957 with the express purpose of monitoring nuclear testing programmes. In the midst of the Cold War, however, the IAEA was beset with difficulties because the nuclear weapon states were wary of allowing international inspectors access to their secret nuclear plants and storage depots.

In spite of the difficulties over monitoring, the nuclear powers agreed to a voluntary moratorium on testing between 1958 and 1961. They also secured the status of Antarctica as the world's first nuclear-free zone; under article V of the 1959 Antarctic Treaty, all the signatories (including the United States, the United Kingdom, the Soviet Union and France) agreed to prohibit all forms of exploding, testing and dumping of nuclear materials. This achievement needs to be tempered with the realization that Antarctica's fate had been secured as a non-nuclear test site at the expense of many other inhabited places in the world. The IAEA had the difficult task of monitoring these testing programmes while introducing systems of audit, report and inspection among the nuclear weapon states. The Vienna-based agency was commissioned to promote the peaceful application of nuclear energy at a time when the superpowers were busy stockpiling weapons and expanding the geographical deployment of these systems.

In spite of these pressures, the international community achieved a measure of progress on arms control through a series of treaties in the 1960s and 1970s:

1963 Limited Test Ban Treaty signed between the United States and the Soviet Union for the purpose of bans on testing in the atmosphere, underwater and in space; only underground testing was permitted

1964 Treaty of Tlatelolco signed by twenty-three Latin American states committing themselves to non-proliferation

1968 Nuclear Non-Proliferation Treaty (NNPT) signed by over one hundred states, including the superpowers, but excluded nuclear weapons nations such as France and China; the treaty was ratified in 1970

1976 Outer space declared a nuclear weapons free zone (NWFZ); treaty signed by 113 states

The NNPT was critical because it provided a benchmark for nuclear proliferation control and strengthened the role of the IAEA in terms of preventing the illict use of nuclear technology. Agreement was reached that a review conference on nuclear usage would be held every five years for the purpose of investigating and assessing the state of nuclear knowledge and the adoption of peaceful uses of nuclear energy. The nuclear non-proliferation and disarmament process was undoubtedly assisted by a number of agreements; here are some of them:

1971 Indian Ocean declared a zone of peace
1972 Nuclear testing on the seabed banned within a 12 nautical mile zone of territorial waters
1974 Threshold Test Ban Treaty between the two superpowers limited testing to a 150-kiloton explosive yield
1985 South Pacific declared a zone of peace
1986 Intermediate Nuclear Forces Treaty between the United States and the Soviet Union
1997 Comprehensive Test Ban (CTB) Treaty; all forms of nuclear testing banned

The attempts to control the spread of nuclear weapons were hampered by the contested politics of the Cold War when, with the help of a number of Western European suppliers like France and Germany, nuclear threshold states such as South Africa, Israel and Argentina developed nuclear programmes. Whereas the United States was anxious not to supply nuclear technologies to the Middle East and Latin America, German and French companies assisted in the construction and production process. It is alleged, for example, that this enabled the South African and Israeli governments to produce a large and unexplained nuclear explosion in the Southern Ocean in 1979. Two years later, Israel launched an illegal bombing raid on Iran's German-made nuclear complex in Bushehr, because of fears that an Iranian nuclear programme would compromise Israeli national security. At the same time, nuclear weapon states issued a number of assurances, so-called negative security assurances, that no nuclear weapons would be used against non-nuclear weapon states. Only China, however, acknowledged at the 1978 UN Session on Disarmament that it would never use nuclear weapons (as a first-strike weapon) on any state unless it was attacked by a nuclear weapon state.

Over the last twenty years, the NNPT has been widely recognized as the major mechanism for global nuclear disarmament. In 1995, the fifth major review conference (held between the participants in New York) was critical because the 1968 treaty stipulated that after a duration of 25 years the participants were entitled to call for a review of the NNPT. A total of 185 members voted to extend the treaty indefinitely; but on a less positive note, calls for a Middle Eastern nuclear-free zone were not endorsed by the treaty parties, in spite of strong support from other countries in regional nuclear-free zones such as Latin America and the South Atlantic. The main sticking point was the refusal of Israel to dismantle her unconfirmed nuclear capability until regional peace was secured, and Iraq's continuation with mass destruction weapon programmes in violation of the UN resolution, which demanded that UN inspection teams should be allowed to verify disarmament.

Notwithstanding the commitment of the NNPT to non-proliferation and disarmament, a number of outstanding issues remain. The first issue concerns compliance and refers to states that have already 'given up' their nuclear weapons, such as Ukraine and South Africa, and those who refuse to accept the authority of the IAEA, such as Iraq and North Korea. In 1993, President de Klerk announced that South Africa had constructed six nuclear devices, which would be dismantled in the near future. On UK television the Channel 4 programme 'Dispatches' alleged that as many as twenty-five devices were created and that the South African nuclear programme had not been totally scrapped. Fears remain that elements of the far right obtained nuclear devices and may seek to blackmail future governments into granting a *volkstadt* (independent state) for the Afrikaaner population. Other states such as North Korea and Iraq continue to cause problems for international inspectors because of their refusal to allow routine examinations of their programmes. The second major concern regards the relationship between the nuclear weapon states and the near nuclear weapon states such as India and Pakistan. They maintain that the United States has been hypocritical in the sense it has allowed Israel to develop a nuclear programme yet endeavoured to restrict the nuclear ambitions of South Asian and Middle Eastern states. Unless and until the world commits itself to complete nuclear disarmament, Indian and other political figures remain reluctant to end their weapons development programmes. Some realist writers, e.g. Kenneth Waltz, have even suggested that further proliferation might actually stabilize world politics if the capacity for nuclear weapons is spread more widely in the international arena.

The nuclear non-proliferation process is intimately linked to nuclear testing. The NNPT conference in New York recognized that advocates in favour of eventual nuclear disarmament had to recognize there was a wider nuclear weapon cycle. The major problem concerning nuclear testing is that treaties signed in the 1970s, such as the Threshold Test Ban Treaty and the Peaceful Nuclear Explosions Treaty, have not been enforced in spite of appeals by the UN Disarmament Commission and the Conference on Disarmament. Both these treaties attempted to restrain test explosions and sought to implement a system of verification among the international community and interested observers. It was only in the aftermath of the Cold War that further progress was possible on nuclear testing. Notwithstanding fears of confrontation in South Asia, major dilemmas for the IAEA and the CTB parties remain: Who is best suited to oversee the ban on all forms of nuclear testing? And what sort of controls or constraints could be imposed by the international community on states which breach the testing ban? Ultimately, the biggest question of all remains, Will the established nuclear weapon states give up their nuclear weapons and commit themselves to permanent disarmament?

Figure 5.2 The defining image of the twentieth century: the nuclear bomb? © Rex Features

Resisting nuclearization: regional initiatives and nuclear criticism

Western powers celebrate the achievements of the 1959 Antarctic Treaty for declaring the Antarctic a nuclear weapons free zone (NWFZ), and while this was a laudable achievement, it failed to acknowledge that the uninhabited polar continent differed enormously from the islands in the South Pacific and the Siberian interior. In 1959 twelve nations signed a landmark treaty in Washington DC which created a legal and political framework for the management of the Antarctic continent and surrounding seas. The major elements of the Antarctic Treaty included establishing scientific activity as the major concern of interested parties; preventing all forms of nuclear testing, dumping and explosions in the region; and creating a forum where the environmental protection of Antarctica would be a priority. Previous territorial claims to the continent were considered suspended for the duration of the treaty. Since its ratification in 1961, forty-three member states have accepted the principles of the Antarctic treaty and subsequent additional measures and protocols (Figure 5.2).

Carol Cohn's ethnographic analysis of strategic culture in the United States demonstrated that the language of technostrategic discourse tended to underplay the dangers of nuclear weapons, and it generated abstractions and euphemisms which failed to connect the deadly potential of bombs to everyday life. During the 1950s, therefore, the Antarctic's empty landscapes simply facilitated this form of strategic discourse, which gave no account of the needs for local people directly affected by particular nuclear strategies (Cohn 1987).

In contrast, a form of extranational discourse based on antinuclearism has been a significant feature of the English-speaking world since the late 1950s. Peace movements such as the Campaign for Nuclear Disarmament (CND) have stressed the global dangers of nuclear weapons and poured scorn on the idea that the bomb has contributed to greater global security (Seager 1993; Pettman 1996). Public demonstrations and marches in the United Kingdom and the United States added to a body of opinion in these societies that nuclear weapons were dangerous and a direct contributor to further global tensions.

In his book *The Control of the Arms Race*, the British scholar Hedley Bull noted that technostrategic discourses on nuclear weapons tended to assume that security referred to the state of superpower relations rather than the wider world (Bull 1965). However, by the 1960s it was abundantly clear that the possibility of nuclear war threatened all citizens, regardless of their location. The development of rocket technology and nuclear delivery systems facilitated global access, as nuclear missiles could theoretically travel across national boundaries. Therefore, talk of security in terms of superpower relations was to marginalize the condition and prospects of the vast body of humanity. It also revealed the profound inequalities between nations; for example, the United States enjoyed considerably more advantages not only in terms of nuclear stockpiles but also in the deployment of surveillance technologies.

Criticism by peace campaigners and academics began to gather momentum in the 1980s. Representatives of CND began demonstrations and a long vigil outside the American airbase at Greenham Common in Berkshire after Mrs Thatcher's decision to allow the deployment of cruise missiles in the United Kingdom as part of the global defence of the free world. For their detractors, women attached to military places such as Greenham Common in the United Kingdom and Pine Gap in Australia were often labelled 'hysterical' and 'mad' by the media reporters and popular newspapers (Cresswell 1996). The response reflected the unease at the sight of militant women and allegedly unfemine behaviour, but more seriously it questioned the wisdom of stationing nuclear weapons in heavily populated areas of the United Kingdom (such as Greenham Common in Berkshire) and challenged the notion that it would enhance the security of the population. In the United Kingdom and the United States, the early 1980s were dominated by the aggressive Cold War militarism of Prime Minister Thatcher and President Reagan, which led to the ever greater deployment of nuclear missiles in Europe.

For peace campaigners, America's development of a Strategic Defense Initiative (Star Wars) in the early 1980s also highlighted the global dangers posed by a nuclear confrontation and the seeming inability of any state, including the United States, to protect its citizens from nuclear missiles. Simultaneously, concerns over nuclear power programmes had escalated following the Three Mile Island nuclear power station emergency in 1979 (United States) and the destruction of the Chernobyl nuclear plant in 1986 (Ukraine). The meltdown of the Chernobyl reactor number 4 resulted in the release of a massive dose of radiation into the local environment, which subsequently precipitated a massive transboundary movement of radioactive material into

Western Europe. Earlier optimism of cheap and safe nuclear energy had been replaced by widespread fear over the safety of power plants and the storage of nuclear waste.

In the South Pacific, the New Zealand government took the unusal step of adopting the demands of antinuclear campaigners by declaring that American nuclear warships were no longer welcome in New Zealand territorial waters. It was a deeply controversial move because New Zealand and the United States had been close Cold War allies under various security arrangements. Nevertheless, in the 1980s the Wellington government began to question the wisdom of collective defence and refused to countenance nuclear weapons on their soil or in 'their' waters. Prime Minister David Lange became a widely cited and admired political leader within the antinuclear movement for his support of a South Pacific Zone of Peace and Cooperation. American irritation was considerable because as John Dorrance, a US consul general in Sydney, noted:

> There are those who fear the dangers of engagement. They are led to an understandable concern about the horrors of war to argue that a policy of non-involvement or isolationism is the way to save our countries from a nuclear holocaust. Unfortunately, we are all now physically within the reach of nuclear weapons and their secondary effects . . . the first requirement must be to maintain our collective strength to deter aggression. (Dorrance 1985: 217)

In contrast, Lange and peace campaigners argued that New Zealand's security was going to be based on developing peaceful relations with its neighbours in the South Pacific rather than involvement in the northern hemisphere dominated Cold War struggles. Indeed, the cover of the 1987 defence review had a map of New Zealand constructed in such a way as to emphasize the proximity of the country to the South Pacific and Antarctica rather than the northern hemisphere.

In the late 1980s, the New Zealand government and NGOs such as Greenpeace began to focus the world's attention on French nuclear testing. In conjunction with the island states of the South Pacific, Greenpeace orchestrated high-profile media campaigns, using their ship *Rainbow Warrior* to interrupt proceedings at the French nuclear test site on Mururoa Atoll. In July 1985 two French secret agents sunk the *Rainbow Warrior* in Auckland harbour; one person died. This led to widespread condemnation and further rallied regional support. The 1987 declaration of a nuclear-free zone in the South Pacific did not deter the French, and in September 1995 Greenpeace relaunched its campaign when France resumed testing in the midst of negotiations for a comprehensive nuclear test ban in Geneva. Local political leaders condemned the French action as geographically and socially insensitive. Sir Geoffrey Henry, prime minister of the Cook Islands, noted:

> I am prepared to accept that by some political concoction, the French have the right to test there, but geographically it is not theirs. It is part of the Pacific. It is as if an invasion has taken place.

Popular protest in the South Pacific culminated in rioting in Papeete, Tahiti (boosting support for the movement for independence in the French colony of New Caledonia), and there were demonstrations in New Zealand and Australia. In 1997 the French government finally announced that it would stop all forms of nuclear testing in the region as part of the NNPT ratification process. For those in the South Pacific, the real costs of nuclear weapons have been the testing process and the long-term effects of nuclear exposure.

In June 1996, at the South Pacific Forum, former New Zealand prime minister Jim Bolger proposed that the entire southern hemisphere should be declared an NWFZ. In doing so, Bolger was echoing a widely held opinion in Australia, New Zealand and the South Pacific that the nuclear weapons process was largely controlled and perpetuated by northern hemisphere nations who happened to test their weapons in the South. In order to advance this goal of a Southern NWFZ, it was decided to work with the United Nations via a range of regional agencies and organizations such as the South Atlantic Zone of Peace and Cooperation, the African Nuclear Weapon Free Zone, the Indian Ocean Rim Initiative and the South Pacific Zone of Peace and Cooperation. The proposal seemed particularly poignant because it was at this meeting of the South Pacific Forum that the Marshall Island government reported on a proposal by the US government to turn several contaminated atolls into repositories for nuclear waste (67 nuclear explosions had been carried out by the United States between 1946 and 1958).

Despite the unequal geographies of nuclear testing, US government officials commissioned new reports into the alleged threats posed by nuclear proliferation in the Third World. After a series of consultations with defence contractors, it was announced in the early 1990s that Lockheed had been awarded a US$689 million contract to develop a theatre high-altitude area defence missile (THAAD) system, which would intercept hostile incoming missiles threatening US targets. At present, the system remains in the development stage, but when completed it is hoped it will provide the kind of protection envisaged by the Star Wars programme initiated by President Reagan in the 1980s. In conjunction with missile programmes, the Clinton administration has also endorsed the continuation of nuclear testing via computer simulation models. Judging by the activities of the United States, proliferation of nuclear weapons remains a major force in world politics.

Nuclear nationalism in the 1990s: India and Pakistan

The non-proliferation of nuclear weapons has been a major priority for post-Cold War US foreign policy. In the midst of the Cold War, the superpowers in 1968 played a key role in formulating a Nuclear Non-Proliferation Treaty (NNPT). In spite of American success in persuading former Soviet republics such as Belarus and Ukraine to give up their nuclear arsenals in 1991 (in return for financial credits and investment), non-proliferation has been a problematic issue not least because three of its old Cold War allies refused to sign the NNPT: India, Pakistan and Israel.

For many supporters of non-proliferation, the NNPT remains the best hope for global disarmament because the leading nuclear powers agreed to abide by common rules: no nuclear weapon state (NWS) may transfer such weapons to other states; no non-NWS may develop these weapons; and the IAEA must check that the NNPT is being properly observed. Two key parts of the NNPT are article IV and article VI. Article IV allows an NWS to develop technology for the peaceful use of nuclear energy, and article VI commits an NWS to pursue effective measures designed to promote global disarmament. Sceptics have argued that the aims of the NNPT, although laudable, are inadequate in terms of dealing with nuclear non-proliferation. According to Joseph Rotblat, winner of the 1995 Nobel peace prize:

> The somewhat ambiguous wording of the Article [article IV] has been cunningly exploited by the nuclear-weapons states to allow them to wriggle out of the obligation it imposes on them: one interpretation is that Article IV commits them only to pursue negotiations; another is that nuclear disarmament should be attempted only as part of a general and complete disarmament. However, the main purpose of the NPT is nuclear disarmament: the Preamble makes this quite clear. All signatories of the NPT accept certain obligations: the non-nuclear weapon states not to acquire nuclear weapons, and the nuclear-weapon states to get rid of theirs. (Cited in Huque 1997: 4)

Some 178 countries signed the NNPT by May 1995, yet the dangers posed by nuclear proliferation are only too apparent in the South Asian region. China, as a major nuclear power, only formally acceded to the treaty in 1992. Since independence in 1947, India and Pakistan have been locked in a bitter political and economic competition, which has extended to territorial conflict and nulcear weapons proliferation. Ten years after the first Chinese nuclear detonation, India was the first South Asian state to emulate this feat by secretly exploding a nuclear device in 1974 at the same time as Pakistan proposed to the United Nations that South Asia should be an NWFZ. Indian officials have consistently argued that the NNPT was attempting to protect a privileged cartel of states which discriminated against non-nuclear Third World states. After the Indian nuclear event, the Pakistani government at the time committed itself to producing a nuclear bomb too, and by 1992 it was confirmed by government officials that Pakistan now possessed a nuclear capability. During this period both countries were enhancing their missile capability; Pakistan developed the Half-2 and India developed the Prithvi missile (Huque 1997: 3). Political rivalry, issues of national prestige and projection of national power were widely held responsible for the massive investment in time and resources.

India and Pakistan consider the NNPT to be discriminatory as it seeks to prevent Third World states from developing technologies enjoyed by the North for the last fifty years. The IAEA has also been accused of pursuing so-called threshold states such as Iraq with inspection visits while ignoring the activities of Israel and other Western allies. As the Indian representative to the United Nations, K.P. Unnikrishnan noted in 1993:

> India could not subscribe to a Treaty or an attitude which divides the world into haves and have-nots, with an inherently inequitable set of responsibilities and obligations of the two. (Cited in Huque 1997: 11)

In the absence of total disarmament, India declared a right to develop nuclear technologies regardless of the NNPT. American attempts to restrict the nuclear development programme in India have enjoyed only moderate success, largely limited to urging the Indian authorities to adhere to IAEA safeguards for nuclear installations. Pakistan has appeared to be more supportive of the NNPT, and since the 1970s it has endorsed the idea of a nuclear-free zone for South Asia. However, both India and Pakistan have refused to sign the NNPT unless it is possible to achieve mutual agreement over nuclear weapons in the the region and a joint signing of the treaty.

The South Asian experience has exposed the real problems faced when trying to persuade sceptical Third World countries to reject nuclear weapons technologies. American approaches to South Asian non-proliferation have often been contradictory: on the one hand, American technological and financial aid enabled both India and Pakistan to develop their weapons capability; and on the other hand, according to Pakistan, American military assistance has favoured India in the shape of joint naval exercises. Moreover, the Americans recently prevented shipment of twenty-eight F-16 jet fighters to Pakistan, aircraft long paid for by the Bhutto government. The prospects for regional disarmament in South Asia would be assisted by a more consistent policy from the United States, a policy that could link together the regional and global disarmament process (Slater, Schultz and Dorr 1993).

The nuclear stakes were raised again in May 1998 when India's government, led by Atal Behari Vajpayee, announced that five nuclear tests had been carried out in the Thar Desert close to the Pakistani border. According to the prime minister, the tests were necessary because India's national security had to be assured in the light of the 'threat' posed by Pakistan and China, and the nuclear states of the Middle East. A formidable community of scientists, defence experts and political commentators gathered in New Dehli to celebrate the testing and to neutralize those who sought to condemn the nuclear demonstrations. In the midst of the celebrations, few media and political commentators showed much concern for the prolonged dangers faced by the tribal population in the uranium mining belt of Jaduguda in Southeast Bihar. National security dictates and the precarious condition of Indo-Pakistani relations are cited as reasons for continuing nuclear tests (W. Walker 1998). During the tense period of May–June 1998, Hindu nationalists seized upon the nuclear testing programme as evidence of India's greatness and proposed that a temple should be constructed to commemorate the site of the five nuclear explosions – a *shaktipeeth*.

As a direct result of these tests, the Pakistani government ordered similiar nuclear testing to be carried out in June 1998. Despite American pleas for restraint, the Pakistani tests were seen by many South Asian observers as a necessary 'show of strength'. The US government and other Western governments, such as Germany, retaliated by imposing sanctions on both India and Pakistan. The underlying problem of Indo-Pakistani relations is the inability of either side to transcend the narrow confines of military security and conceive a form of human security which trangresses borders and particular national spaces. Nuclear weapons testing therefore becomes just another element in a dangerous and escalating struggle over disputed territories in the Kashmir and the Punjab, which have witnessed further massacres and shelling in the middle of 1998.

More generally, American attempts at nuclear reduction have only been moderately successful, and Washington's support for US–Russian joint reductions in 1991 and 1993 have to be counterbalanced with a reluctance among the former superpowers to agree to a long-term ban of all forms of nuclear testing. In 1995 the Americans proposed that even after the ratification of the Comprehensive Test Ban Treaty (negotiations began in 1994), there should be an opt-out clause after the first review conference in ten years' time. This was widely condemned and in 1997 the treaty was endorsed by all countries, including other high-profile nuclear testers such as France and China. For South Asian observers the reluctance of the Americans to commit themselves to total nuclear disarmament leaves the way open for others to follow suit, until the original superpowers make the final commitment (Ayoob 1993, 1995).

Conclusion

This chapter has explicitly rejected the claim by some conservative commentators that nuclear proliferation may actually contribute to regional and global stability because it introduces a sense of caution into the decision-making process of political and military leaders. Nuclear weapons also raise fundamentally moral issues, and these often conflict with realist political views that stress national security and political self-interest (White, Little and Smith 1997). The persistence of nuclear weapons remains a striking symbol of the globalization of danger in the sense that a number of countries possess weapons of mass destruction. In the aftermath of the Cold War, these fears over nuclear, biological and chemical weapons have not evaporated, as recent events testify in India, Pakistan and Iraq.

Critical geopolitical writers such as Simon Dalby have argued that nuclear weapons have to be located within a matrix which acknowledges that militarization is intimately connected with other concerns such as North–South relations, development, poverty and environmental problems (Dalby 1991). Promoting processes of demilitarization requires us not only to consider how nuclear weapons can be eradicated but also why states seek to retain these weapons of mass destruction. Any future management of international nuclear behaviour will therefore require a careful analysis of global security, and it must be sensitive to existing levels of wealth, consumption and environmental degradation (Commission on Global Governance 1995; Burchill and Linklater 1996).

Further reading

Have a look at R.J. Lifton and R. Falk's *Indefensible Weapons: The Political and Psychological Cases Against Nuclearism* (Basic Books, 1982) and J. Newhouse's *The Nuclear Age* (Michael Joseph, 1989). M. Huque has published an interesting article 'Nuclear proliferation in South Asia and US policy', *International Studies* **34**: 1–14. For feminist arguments concerning nuclear weapons see J. Seager's *Earth Follies* (Earthscan, 1993) and J. Pettman's *Worlding Women* (Routledge, 1996). And for critical geopolitical evaluations see S. Dalby's *Creating the Second Cold War* (Belhaven Press, 1990).

Chapter 6

The globalization of environmental issues

On the eve of Operation Desert Storm, Prince Hassan of Jordan noted:

> If the oil wars have begun, the water wars may not be far behind and then conflicts stemming from the consequences of soil loss, foresty loss, high global temperature and rising sea levels will follow. (Cited in Piel 1992: 311)

For some environmentalists and academic commentators, Prince Hassan's comments would not be considered unduly sceptical in the face of rising population growth, resource exploitation, widespread poverty, airborne pollution, ozone depletion, coastal and oceanic pollution, forestry loss, fossil fuel consumption, and new information concerning the human impact on the environment (Palmer 1992: 182). There is now a widespread sense in which the limits of social progress and industrial development are being reached, hence many realist writers fear that future international environmental relations will be dominated by disruption and violence as states seek to preserve their environments or their resource access (Princern and Finger 1994; Doyle 1998).

The ending of the Cold War has promoted some positive momentum in the northern polar region and led to the creation of the Arctic Council in 1995 after six years of painfully slow negotiation. In 1989 the Canadian government proposed to eight other Arctic states that a regional council should be created to formulate joint action strategies for the environment, development and cultural preservation, and that indigenous people's movements such as the Saami Council and the Association of Indigenous Minorities of the North should be invited to join this council. After overcoming the initial opposition from the United States, the Arctic Council faced the difficult task of combining a commitment to the Arctic Environmental Protection Strategy with economic development for marginal areas. A classic case was the decision by the Canadian government to approve a diamond mine in North West Territories in August 1996, in spite of ecological criticism. Most governments involved in the council appear to lack the commitment to enforce environmental protection when faced with economic pressures from either indigenous peoples or settlers (Chaturvedi 1996).

More generally, environmental and developmental issues have assumed a far higher political profile (Imber and Vogler 1995). Growing concern over environmental degradation and the possible impact of climatic change has been linked to contemporary debates over planetary security, militarism and

Figure 6.1 During the 1982 Falklands/Malvinas War, the Argentine forces planted over thirty thousand mines on East Falkland. Although many mines have been destroyed, the island is still littered with reminders in the shape of warning signs. (*Photo*: Klaus Dodds)

global governance (Figure 6.1). This concern for environmental affairs within the arena of global politics was firmly consolidated at the United Nations Conference on Environment and Development (Earth Summit) held in Rio in June 1992. Hundreds of NGOs were joined by more than a hundred governments and heads of state to discuss the numerous dangers facing the global ecosystem. Concern for environmental change has a long history, as worldwide industrialization and urban development have given rise to new perceptions about the earth as a single interconnected biosphere.

Most writers concerned with environmental affairs recognize that ozone depletion, climate change and environmental degradation pose troubling questions for global politics; this is because they are often transboundary issues and are therefore beyond the remit of any one state (Shafer and Murphy 1998). Human survival depends on the capacity of states and other organizations to collaborate in unprecedented ways in order to protect the earth's ecosystems. Others are more hopeful by recognizing that the fate of local ecosystems is in part tied to transnational systems of production and exchange. Globalization theorists and Third World writers continue to emphasize that environmental issues have to be considered as part of a wider matrix concerning poverty, consumption, development and North–South relations (Mittleman 1996; Spybey 1996; Thomas and Wilkins 1997). In its 1998 report on human development, the United Nations recognized that the world's poorest 20 per cent enjoy only 1.1 per cent of the world's income (United Nations 1998). Whereas Northern countries worry about climate change and ozone depletion, Southern states are often more preoccupied with population increase, resource scarcity, basic needs and poverty reduction. Hence our definitions of 'environmental issues' can often determine the sorts of analyses and policy options we ultimately derive (Lipschultz and Conca 1993; Hurrell 1995).

This chapter reviews the growing significance of environmental issues in world politics. The first section charts the rising profile of environmentalism from the 1960s onwards. Next comes the Rio Earth Summit, considered in some detail with particular reference to debates over global governance and environmental security (Middleton, O'Keefe and Mayo 1993). The third section focuses on the international community's attempts to manage the global commons of Antarctica, the atmosphere and the ocean floors. The final section examines the ongoing tensions created by environmental issues in world politics. It suggests that a managerial approach to the global environment needs to be supplemented and then replaced by an ecological approach which recognizes that our relationship with the world around us will have to change.

From Stockholm to Rio: transboundary and global agendas

Environmental concern has emerged as a significant feature of global politics since the late 1960s. *Silent Spring*, the best-selling book by Rachel Carson, helped to mobilize public interest in environmental politics (Miller 1995: 6). More specifically, the creation of movements such as Greenpeace and Friends of the Earth in 1969 consolidated interest in conservation throughout North America, Australasia and Western Europe. Some environmental thinkers such as Ronald Inglehart have argued that this concern was also indicative of a new generation in post-war Europe and North America anxious to warn the wider world of the dangers of modern industrial development and nuclear fallout (Dobson 1990). Media coverage of environmental disasters such as the sinking of the oil tanker *Torrey Canyon* – 875,000 barrels of crude oil were later washed ashore along sections of the Cornish coast – alerted people to the environmental dangers posed by massive oil spillages (Figure 6.2).

Greepeace was founded in the Canadian province of British Columbia in 1971, initially to oppose the US underwater nuclear testing in Alaska. The group's reputation for direct action was cemented in that campaign, after

Figure 6.2 Prince William Sound, Alaska: clean-up following the *Exxon Valdez* oil spill. (*Photo*: Heidi Bradner/Panos Pictures. © Heidi Bradner 1993)

sailing their vessel into the centre of the test site. Although unable to prevent the nuclear tests, by the mid-1980s the group had successfully transformed a loose coalition of environmental writers, academics and sailors into a multinational organization. After the US government agreed to end underwater testing in 1972, the prevention of all further testing became the new Greenpeace priority. Their direct action helped to achieve the following outcomes:

- French nuclear testing suspended in the South Pacific in 1975. When France renewed testing in the 1980s and 1990s, Greenpeace mobilized opposition and continued to campaign even after their vessel, *Rainbow Warrior*, was sunk in Auckland harbour in 1985.
- The International Whaling Commission announced a moratorium on whaling in 1982 following Greenpeace's protests against the practice.
- The 1983 London Dumping Convention established a moratorium on ocean dumping of radioactive waste following a Greenpeace campaign.
- In 1987 Greenpeace established an Antarctic base in order to highlight their opposition to all proposals which would allow mining in the region. A new environmental protocol was established for Antarctica in 1991.
- In 1995 Greenpeace prevented Shell from dumping the Brent Spar oil platform in the North Sea. The oil rig was later towed to Norway for dismantling on land. Disputes over the scientific data concerning the environmental impact of sea dumping, lost Greenpeace a degree of credibility.
- Through their direct action and skilful use of the media, Greenpeace have become one of the most effective environmental organizations, with group membership in more than twenty-five countries and public subscriptions and donations of over £30 million pounds a year.

Since the mid-1960s, debates on international and/or transboundary pollution have occupied environmentalists and governments in the North. There was concern that the effects of industrial pollution were placing not only dangerous pressures on state cooperation, but also that the ecological limits of the earth had been reached. The **transboundary** nature of phenomena such as acid rain made it evident that local and regional environments could not be maintained or even protected by individual states or regional strategies. The term 'acid rain' was adopted by the media and enviromental campaigners in the 1980s to highlight growing awareness of airborne chemical pollution (in the form of sulphur dioxide and other pollutants) and its subsequent effect on vegetation and the food chain. While scientists have known about the problems of industrial pollution for some time, it was only in the last decade that political leaders started to press for preventive action. Prime Minister Thatcher acknowledged in April 1986 that UK industrial pollution was causing acid rain to fall on Scandinavian forests with devastating consequences. Political scientists and geographers call this transboundary pollution because acid rain can fall on areas many miles away from its original source, such as factories, power stations and volcanoes. International agreement concerning the reduction of acid rain has been slow, with some states such as the United Kingdom and the United States unwilling to accept large-scale reductions called for by the Nordic countries.

This necessitated a shift in attitude from local problems of pollution and waste management to global issues of ecosystem management. In 1968 the Intergovernmental Conference of Experts on the Scientific Basis for Rational Use and Conservation of the Resources of the Biosphere met in Paris to discuss human impact on the biosphere, including issues such as overgrazing, deforestation and water pollution. The scale of human activity is such that many 'green' commentators and scientists are concerned that the maximum capacity of the biosphere to absorb and sustain these activities is rapidly reaching saturation point. High-profile reports such as the *Limits to Growth* (1972) claimed that economic growth, in terms of increased production and consumption of goods and services, could not continue in an unchecked manner, because of the implications for future environmental management. The metaphors employed by the neo-Malthusian writers of the 1970s portrayed the earth as a lifeboat and/or as a spaceship to highlight its ecological capacity to handle pollution and economic growth. While this tranche of literature was often criticized and condemned for failing to acknowledge North–South inequalities and the social mechanics of life, it did acknowledge the limits to development (Doyle 1998; Doyle and McEachern 1998).

A UN-sponsored conference on human development took place in Stockholm in 1972. It attempted to advance some of the themes addressed at the 1968 intergovernmental meeting in Paris. In contrast to the Paris meeting, the Stockholm conference embraced the political, economic and social issues connected with human development and it attracted states and non-governmental organizations (NGOs). Lobbying by the Third World political coalition, the G77, ensured that the Stockholm agenda considered issues of particular interest to them, such as water supply, poverty and shelter; and this broadened the discussion beyond Northern concerns over population growth, resource exploitation and limits to economic growth (Miller 1995: 8; Williams 1997). The G77 also called for an acknowledgement of the links between welfare, industrial development and environmental degradation. For the first time in a public forum, differences between North and South over environmental issues became abundantly obvious (Chapter 3). Three major issues raised by G77 delegates were: Who was responsible for environmental degradation? Did the South have a right to develop along the same lines as the North? And should the North offer the free transfer of 'clean technology' to the South?

After the 1972 Stockholm Conference on the Human Environment, the UN adopted a more active role by furthering ozone layer protection (Montreal Protocol 1987), regulating the disposal of hazardous waste (Basel Convention 1989), highlighting the problem of overfishing through producing reports on the world's fisheries by the Food and Agricultural Organization (FAO) and limiting further tropical deforestation through the Tropical Forest Action Plan. The United Nations Environment Programme (UNEP) was created following the 1972 conference; its purpose was to address global environmental issues and North–South relations. Over the next twenty years, environmental and economic issues were debated ranging from ozone depletion, forestry and sustainable development to climate change. Northern states such as the United States have concentrated their diplomatic energies on ozone depletion, whereas Southern states such as India have sought to draw wider connections between environmental destruction, poverty, debt and development. Although the G77 had a limited impact on the outcome of these negotiations, it provided a forum for Southern accusations that Northern states were reluctant to acknowledge how most of the human-induced damage to the earth's biosphere was the result of their industrial activities and/or that Northern states were now trying to curtail the developmental aspirations of Southern states. The Indian writer Vandana Shiva referred to this situation as a form of 'ecological imperialism', whereby the North seeks to instruct the South on how to reform its industrial behaviour while refusing to assist in debt reduction and/or technology transfer (Shiva 1993).

The linkage between North–South relations and environmental issues was recognized in the Brundtland Report of 1987. By using the term 'sustainable development', it was explicitly acknowledged that the developmental needs of the South could not be marginalized by the Northern agendas of global environmental protection. However, throughout the 1980s, the Northern bloc of the United States, Europe and Japan were unwilling to make concessions over industrial development and the consumption of resources; and the South, led by countries such as India, China and Brazil, refused to relinquish control over their right to determine their development priorities in the face of evidence that the North (25 per cent of the world's population) consumed 70 per cent of the world's energy, 75 per cent of the world's metals and 60 per cent of total global food production. The problem facing many of the negotiators at major international conferences over climate change, ozone depletion and global warming was that no consensus could be reached over the nature of environmental issues, the meaning of sustainability and the core principles of management for the future. During the 1980s, therefore, environmental groups argued that states frequently committed themselves to non-binding conventions which respected their sovereign interests at the expense of developing global and politically inclusive forms of protection for the environment.

Under the guise of sustainable development, environmentalists have argued that major states such as the United States and multinational corporations have sought to project a particular vision of sustainable development, a vision which privileges the capacity of the market and industrial development to produce ecologically friendly economic growth (Redclift 1987). In doing so, the limits to growth arguments have been dispensed with because the increased efficiency of industrial farming and other production systems can effectively bypass these ecological limits. Larry Summer, a World Bank economist, made the following observation in 1991:

> There are . . . no limits to the carrying capacity of the earth that are likely to bind any time in the foreseeable future. There is not a risk of any apocalypse due to global warming or anything else. The idea that we should put limits to growth, because of some natural limit, is a profound error. (Cited in Seager 1993: 134)

For ecologists and globalization theorists, this unproblematic vision of the future is deeply troubling because it reduces environmental issues to questions of efficiency and effective planning, rather than recognizing that some profound moral and political issues are raised by industrial development and economic growth. As Michael Redclift noted in his often cited critique:

> Sustainable development, if it is to be an alternative to unsustainable development, should imply a break with the linear model of growth and accumulation that ultimately undermines the planet's life support systems. Development is too closely associated in our minds with what has occurred in western capitalist societies in the past, and a handful of peripheral capitalist societies today. (Redclift 1987: 4)

The 1992 Rio summit was intended to be a forum for discussion and debate of these controversies. It was also intended to act as a focus for moral pressure on governments across the globe while strengthening the ongoing work of local NGOs, women's groups and community-based organizations (CBOs).

The Rio summit and global ecology

The 1992 Rio summit held on the twentieth anniversary of the Stockholm Conference was attended by 170 states, thousands of NGOs and many multinational corporations. The purpose of the conference was to consider the environmental consequences of human development. Five years earlier, the 1987 Brundtland Report had warned that traditional patterns of economic growth were not sustainable in the long term, given the demands of the South for further industrial development. What became apparent at the Rio summit was that Northern and Southern states were pursuing different environmental agendas: the Northern states were concerned with ozone depletion and global warming whereas the Southern states were anxious to address the relationship between economic development and environmental management. The Rio summit produced conventions dealing with climate change, biodiversity, forestry and Agenda 21, but left unresolved considerable differences and difficulties over the design and implementation of sustainable forms of development.

The main achievement of the Rio summit was to convene a global forum for the discussion of global environmental problems in the wake of UN resolution 44/228, which called for such a meeting in 1989. After weeks of preparatory meetings, the governments agreed to the following: twenty-seven core principles of development and the environment; conventions on biodiversity and climate change; Agenda 21 and a host of other environmental agreements such as the creation of a Commission on Sustainable Development to help the UN monitor environmental progress. The main document, Agenda 21, declared in article I that human beings were central to sustainable development. Article II, however, reiterates that states enjoy the right to exploit their own resources. Underlying this document is a powerful commitment to upholding the right of states to decide their own environmental strategies, even though it is acknowledged that states should seek to act in a sustainable manner. Indeed article 15, the so-called precautionary principle, urged that:

> In order to protect the environment, the precautionary approach shall be widely applied by states according to their capabilities. Where there are threats of serious or irreversible damage, lack of scientific uncertainty shall not be used as a reason for postponing cost-effective measures to protect environmental degradation.

Although this sounds laudable, it trivializes the difficulty of balancing the future needs of humanity with the sovereign interests of states and the business interests of multinationals when the efficacy of scientific information is being challenged in an attempt to halt binding environmental agreements relating to industrial development.

In his critique of the Rio summit, Tim Doyle argues that the key document concerned with sustainable development (Agenda 21) was explicitly enframed by Northern political elites and transnational corporations (Doyle 1998; Doyle and McEachern 1998). While environmental problems were defined in terms of *global ecology*, the problems of global warming and population growth were frequently discussed in Northern elite and scientific terms that marginalized the major environmental issues defined by the people and states of the South. Doyle argues that Agenda 21 effectively perpetuated a form of sustainable development which continues to promote the goals of economic growth and industrial development through market liberalization and world economic regulation. The environment is effectively considered to be a resource that can be used efficiently by particular human 'users', rather than viewing human beings as intimately bound up with the fate of the earth's ecosytems and its other inhabitants. Instead of promoting profound change in human behaviour, the Rio summit effectively approved existing forms of industrial development and outlined an approach for piecemeal change and legislation.

A sense of anger and disappointment with the Rio summit was keenly felt by Third World countries, who noted that Northern states were not willing to alter existing global systems of trade, finance and debt collection. In 1994 a Global Conference on the Sustainable Development of Small Island Developing States was held in Barbados. Representatives from over a hundred states attended the conference to consider the economic and environmental problems faced by small island states. The delegates approved a programme of action which called for measures to protect these states from rising sea levels, the loss of natural resources and the dependency on a few primary exports.

The conference revealed that island states such as the Maldives in the Indian Ocean have very different developmental and environmental agendas to geographical giants such as China and India. The Maldives could disappear if sea levels were to rise in the next century due to ice cap melting, hence agreements concerning global warming are of particular interest, which further acknowledges that not all states and communities share the same priorities (Chaturvedi 1998).

Environmental issues and sustainable development need to be considered in alliance with negative equity and net resource flows from South to North. The Indian ecological writer, Vandana Shiva noted that:

> The 'global' in the dominant discourse is the political space in which a particular dominant local seeks global control, and frees itself of local, national and international restraints. The global does not represent the universal human interest, it

> represents a particular local and parochial interest which has been globalised through the scope of its reach. The seven most powerful countries, the G-7, dictate global affairs, but the interests that guide them remain narrow, local and parochial. (Shiva 1993: 149–50)

Disappointingly, the Rio summit failed to address some of the most pressing problems facing global environmental politics such as securing firm and binding commitments to cutting carbon dioxide emissions, reversing the militarization of the environment and exerting firm controls on the activities of multinationals. As with most of the conventions negotiated at Rio, the Climate Change Convention was replete with ambiguities, omissions and qualifications to allegedly binding agreements. The problem of Third World debt and its linkage to poverty and maldevelopment was also not considered, even though the alternative Global Forum had called for a greater willingness on the part of Northern states and banks to grant substantial debt relief and promote the involvement of non-state organizations in the production of key documents such as Agenda 21.

The trend to privilege the role of the state and the interests of Northern multinational organizations continues in the wake of Rio. In July 1996 the United Nations Development Programme (UNDP) issued another report on sustainable development which actually called for *further* industrial growth in order to tackle the inequalities between North and South. New organizations such as the **World Trade Organization (WTO)** and the Multilateral Agreement on Investment (MAI) are designed to remove any impediments to the global market economy and hence promote further economic expansion. The MAI was negotiated between 1995 and 1998; it will legally bind states and limit their power to impose conditions and requirements on multinational investors, further reducing their leeway to control trade and investment flows in their national territories. Unfettered and open economic growth may well contribute to further environmental degradation and could effectively weaken the capacity of states or non-state movements to counteract or even protect specific environments (Herod, Ó Tuathail and Roberts 1998; Pilger 1998).

In the future it might be necessary to consider new ways to combat environmental challenges instead of abdicating this responsibility soley to nation states. This is not intended to demean the achievements secured by states, such as the Montreal Protocol in 1987 and/or the creation of organizations such as the Commission of Sustainable Development and the Global Environmental Facility. Compared with thirty years ago, many more states are committed to sustainable forms of development, if only in principle, but the fate of areas such as the Kibale forest region in Uganda, which has witnessed massive deforestation and village displacement, is a truly *global* concern (Figure 6.3). The interaction of national governments, transnational corporations, international agreements and international organizations such as the World Bank and the World Trade Organization have implications for all our environments.

Figure 6.3 Fragile semi-arid environments present particular challenges for local people in West Pokot, northwest Kenya. This picture was taken shortly after the onset of the rainy season; notice the gullying in the foreground. In spite of growing population pressures and ongoing land disputes with the Turkana tribal peoples, the Pokot have devised irrigation and soil conservation measures to help protect the local environment. (*Photo*: Klaus Dodds)

North–South relations and protection of the global commons

The enviromental protection of global commons such as the earth's atmosphere is a problematic venture due to the limits of interstate cooperation and the North–South division. Areas of the world which are not claimed by any one nation state are know as *res communis humanitas*; they include the earth's atmosphere, Antarctica, the ocean floors and outer space. The protection of these areas raises questions about the responsibility of the present generation not only for the global environment but also for future human populations.

The notion of a 'global common' and/or a 'common heritage of mankind' has been employed by the international community to signify the regions which are not only outside the sovereign jurisdiction of the state but are also areas that provoke common managerial concerns. The question of responsibility for these areas remains undecided, given that the sovereign rights of states extend to the margins of the Antarctic, territorial waters and airspace (Vogler 1995; Vogler and Imber 1995).

The advance in technological and scientific abilities to exploit and degrade environments, such as the ocean floors and outer space, became increasingly politicized after the 1950s. The First UN Conference on the Law of the Sea established rights for coastal states to declare exclusive rights over the adjacent continental shelf. The ownership of the ocean floor excited much international debate when the significance of these deliberations in relation to fishing and commerce became evident. New technologies such as oil and gas drilling, coupled with the effects of marine pollution, called for an added sense of urgency. There was therefore a widespread awareness that the oceans and seas were now even more vulnerable to the development of international economic enterprises.

The Third United Nations Conference on the Law of the Sea in 1982 had extended this process of state resource rights by including new privileges to the resources of the continental margin (Vogler 1995). By 1992, after thirty years of negotiation, 154 states had agreed that the ocean floors and the sea could be incorporated into state ownership on the basis of a declaration of a territorial sea (up to 12 miles from the coastline) and/or an exclusive economic zone (up to 200 miles from the coastline). Each coastal state could claim, under the 1982 UNCLOS III Convention, a 200-mile zone for the purpose of exploration, exploitation, conservation and management of resources in the sea, seabed and subsoil (Figure 6.4). This process of delimiting ownership of the waters and oceans has been highly unequal because more than half of the world's exclusive economic zones (EEZs) belong to ten countries, including most of the Northern states, such as the United States, the United Kingdom, Japan, Russia and Canada. The biggest gainers in terms of submarine petroleum rights were Norway, the United Kingdom, the United States, Russia and Australia.

The Law of the Sea appears to favour a select number of Northern states, but the enforcement of these rights has become increasingly problematic. In the 1990s, the Patagonian toothfish has become one of the most overharvested fish in the Southern Ocean (Lilley 1997; Dodds 1999). Overfishing in this massive oceanic zone has been a problem in the past, resulting in the severe reduction in species numbers of the marbled rockcod and icefish. The Antarctic Treaty parties (through the Convention for the Conservation of Antarctic Marine Living Resources, in force since 1982) have attempted to regulate fishing, but they failed to prevent illegal fishing in and around the various EEZs of Southern Ocean islands such as Crozet, Heard and Prince Edward. The toothfish, highly valued by American and Japanese consumers and fishing companies, have been harvested at an alarming rate. The sheer expanse of ocean makes it difficult for international bodies and national governments to

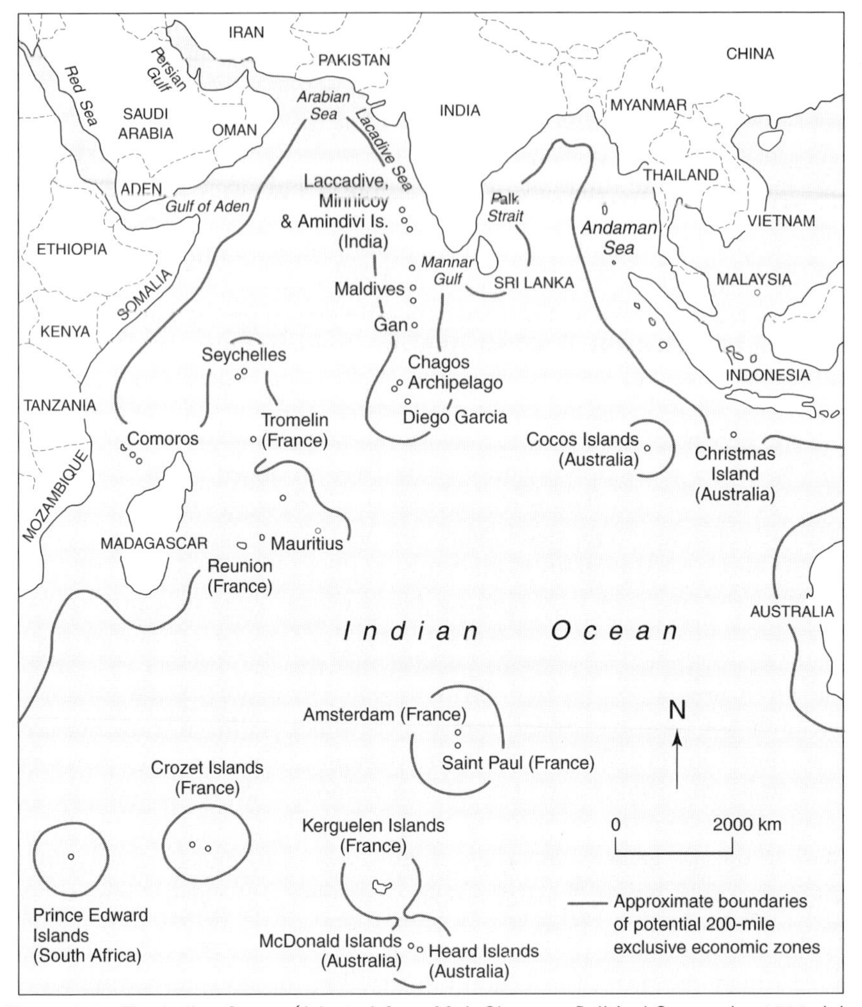

Figure 6.4 The Indian Ocean. (Adapted from M. I. Glassner, *Political Geography*, 1996, John Wiley: Chicester)

regulate fishing and/or conserve fish stocks. Countries such as the United Kingdom, Australia and France have sought to clamp down on illegal fishing activities by arresting vessels and fining their owners (Figure 6.5).

In thc last few years, the attempts to manage fishing in the Southern Ocean have been circumvented by fleets of highly modernized vessels from Spain, South Korea and Norway. A number of claimant states such as France, the United Kingdom and Australia have despatched naval patrols in an attempt to protect their fishing resources within particular EEZs. French patrol vessles captured and impounded a number of vessels around their sub-Antarctic island of Crozet; but for countries such as South Africa, which claims a group of islands in the Southern Ocean called the Prince Edward Islands, limited resources prevent the protection of toothfish stocks. It was estimated that in

Figure 6.5 This Russian fishing vessel is registered in Belize; it is moored in Lyttleton Harbour, Christchurch, New Zealand. (*Photo*: Klaus Dodds)

1996–1997, 30,000 tons of toothfish were taken illegally from South African waters by over thirty fishing fleets. The enforcement of maritime rights in regions as vast as the Southern Ocean can only be piecemeal as fishing fleets and governments expose and exploit the regulatory and surveillance powers of other states and intergovernmental organizations. Open access to the resources of the high seas remains a major problem in terms of common heritage management because no one state or international body can control the movement of vessels and activities in these maritime regions.

The most significant aspect of the 1982 convention was the recognition of landlocked states like Bolivia and Mali to access areas of the ocean, such as the deep seabed, as a common heritage of mankind. Drawing upon the earliest ideas of Arvid Pravo, then Maltese foreign minister, UNCLOS established the ocean floors and their resources as common heritage, decreeing that any resource revenue derived from this area would have to be shared among the international community, regardless of which country exploited the seabed. Simultaneously, the International Seabed Authority (ISA) was created to assist with the process of mining licensing, technology transfers and revenue redistribution. The establishment of the ISA was opposed by a number of Northern states such as the United States and the United Kingdom; they were unwilling to accept the idea of the ocean floor as a common heritage but adamant that the resources of the ocean floor should be available only to those who were prepared to invest in the exploitation.

The management of the global commons highlights issues of interdependence, vulnerability, economic and political justice. Throughout the last thirty years, calls for global environmental protection have had to coexist with demands for a new international economic order (NIEO) and territorial and resource sovereignties, even in areas where claims to sovereignty, such as in outer space, would appear to be ridiculous. The history of outer space as a global common bears witness to the observation that the management of these areas is thoroughly infected with North–South inequalities. As with the ocean floors and the earth's atmosphere, the crucial issue is technology and industrial development. Note that the benefits of exploitative activities are highly unequal, either for the present or future generations, and remember that it was representatives of the South, not the North, who pressed for concepts such as common heritage to be applied to the global commons (Chapter 3). Over the last thirty years, clashes over the exploitation of resources, the usage of satellites and the emission levels of gases have been severe enough to suggest that cooperation has been restricted rather than substantially enlargened since the 1950s.

Cooperation over the global commons will remain deeply problematic in spite of path-breaking agreements such as the 1959 Antarctic Treaty, the 1987 Montreal Protocol and the 1991 Protocol on Environmental Protection (Paterson 1996). One of the enduring problems affecting all ecological issues resides in the difficulty of devising common environmental agendas on the basis of vague scientific evidence and time frames. In that sense the discovery of the ozone hole over the Antarctic in the mid-1980s revealed the opportunities and dangers inherent in tackling global environmental change (Stokke and Vidas 1996; Dodds 1997).

The destruction of the stratospheric ozone layer was closely monitored by the British Antarctic Survey in the 1980s, based on fears that the release of chlorofluorocarbons (CFCs) was responsible for ozone reduction in the polar vortex over the Antarctic continent. As a region far removed from economic activity and population centres, the Antarctic was considered to offer an early warning of impending global damage to the environment, and in 1988, prompted by this evidence from the Antarctic, the World Meteorological Organization and the United Nations Environment Programme set up the Intergovernmental Panel on Climate Change (IPCC) as a discussion forum to exchange scientific information on global warming.

The ethical challenges posed by the global commons were intensified after the 1992 Rio summit, which, following the advice of the IPCC, created a framework convention on climate change (Paterson 1996; Bush and Harvey 1997). While there was agreement that the world faced a series of environmental crises, there was less consensus on the construction and responsibility for joint action programmes. The Climate Change Convention was undoubtedly weakened by the unwillingness of the United States to support a technological and financial transfer to the South, in order to promote clean technologies for industrial development. The American administration also campaigned for a policy on greenhouse gas emissions, which allowed for reduction trade-offs, thereby undermining earlier commitments to the Montreal Protocol in 1987. Some states have clearly been more proactive than others in tackling climatic change. Low-lying countries in particular, such as the Comoros

Islands and the Maldives (part of the Alliance of Small Island States), have a pressing interest in globally binding agreements to reduce greenhouse gas emission.

Since the 1992 Rio summit, progress has remained slow in terms of developing binding environmental conventions to deal with climatic change and biodiversity. The position is not likely to be improved rapidly, as environmental changes often appear gradually and because governments and other interested parties are apt to question available scientific evidence (Vogler 1995; Graham 1996). The Global Climate Coalition, funded by the oil, natural gas and car lobbies, challenges existing evidence relating to global warming patterns and warns that energy bills would rise considerably if measures were taken to reduce carbon dioxide emissions. Major events such as the Basel Convention on Climatic Change eventually agreed and formulated a resolution on the basis of what might be politically possible rather than what might be needed in order to cut pollution levels and halt the exploitation of the environment.

In spite of efforts to reform the consumption patterns of the North, such as the banning of CFC-composed products, the real battleground in terms of global environmental change is likely to be between North and South. Andrew Hurrell has noted how global environmental change

> is an inherently global issue both because of the high levels of economic interdependence that exist within many parts of the world economy and because it raises fundamental questions concerning the distribution of wealth, power and resources between North and South. (Hurrell 1995: 131)

For the South, major economic nations such as China, India and Indonesia argue that their developmental priorities have to be balanced with all calls for global environmental protection. Successive Chinese governments have been deeply suspicious of what they perceive to be the North's environmental agenda at a time when China's economy is growing in output and foreign direct investment. These large Southern economies have therefore been unwilling to compromise their comparative advantage, although often with dire consequences for the environment. Moreover, Northern multinationals may also be partially responsible for Southern environmental degradation. The forest fires in Indonesia during 1997 appear to have been started as a result of commercial logging activities.

This fundamental clash between North and South became evident once again during the negotiations that led up to the Kyoto summit on global warming in December 1997, and it resurfaced in Buenos Aires during November 1998. The Clinton administration was accused of failing to lead the way, by its reluctance to commit the United States to a binding reduction in emission rates (20 per cent of the world's total carbon dioxide emissions) in a set period of time. This prompted the G77 bloc to reject earlier calls in September 1997 to establish binding targets for the South. Oil producers, such as Saudi Arabia, have also demanded that Southern states dependent on oil and natural gas exports should receive compensation from the international community. This move was bitterly opposed by the United States, and it highlights yet again the demands akin to a new international environmental order (NIEO), which recognizes that the North should be prepared to compensate and support the consequences of global environmental reforms in the South.

The fragmentation of Northern consensus on global environmental issues revealed fundamental differences over global warming between the US government and the European Union. The European Union has argued that the North will have to demonstrate a firm commitment to emission reduction before persuading the South to follow suit. EU governments have accepted that the roots of many of the world's environmental problems lie with the North rather than the South. In October 1997 the European Commission announced that EU emission levels for the year 2000 would be 15 per cent below 1990 figures. The United Kingdom's shift from coal-fired to gas-fired power stations in the 1980s and 1990s has contributed to this process by allowing it to claim that a six per cent decline in greenhouse emissions will be achieved by the year 2000. But EU plans to further reduce European levels of emission will be tied to agreements with Japan and the United States, committing these countries to achieve similiar reductions. As in the case of the 1992 Rio summit, the US government remains reluctant (there was agreement in Buenos Aires in November 1998 from the US delegation to cut emission rates) to establish firm target figures in the light of lobbying from industrial and congressional sources which has urged President Clinton not to threaten US jobs and profit margins.

Ultimately, any proposals for the protection of the earth's atmosphere and Antarctica have to recognize the unequal power relations between North and South. Global agreements such as the Climate Change Convention are often flawed in the sense that proposals to cut emissions reflect Western scientific assessments and environmental values (Gupta 1997). In other contexts, the environmental protection of the Antarctic is determined primarily by the United States, the United Kingdom, New Zealand and Australia, thereby marginalizing Southern concerns about rules and regulations for these regions. Northern NGOs such as Greenpeace have therefore been important advocates of more precautionary action and more appropriate behaviour in the wake of these ever widening North–South inequalities.

Conclusion

In October 1998, British newspapers reported that Greenpeace was involved with the UK food company Iceland in the production of new ozone-friendly freezers and fridges called the Kyoto range. The name refers to the city that hosted the summit on global warming in December 1997. In an unprecedented move, Greenpeace agreed to endorse this range of appliances and this undoubtedly helped Iceland's promotional campaign. Interestingly, most of the UK newspapers carrying the story used images of the Antarctic to indicate that fears over ice cap melting have intensified commercial pressures to develop ever more environmentally friendly technology. Place-based images, environmental and political issues are bound together in a complicated manner.

Global politics related to environmental concerns calls into question the capacity of states to deal with climate warming and other global problems; the ability of the international arena to facilitate cooperation; the willingness of environmental movements and transnational corporations to propagate regionally sensitive policy options and strategies; and the role of international institutions and regimes in contributing to a wider culture of obligation. The inability of international regimes to sanction action against states and other organizations which fail to meet particular environmental standards is a worrying problem. Who will punish China or the United States, for instance, if they fail to adhere to their carbon dioxide emission quotas? The answer may lie in a coalition of states and organizations, including NGOs equipped to exert pressure and adept at shaming parties who fail to meet their international obligations. Television and other media networks might further environmental action by exposing wrongdoing on the part of states and multinational corporations.

Environmental issues remain a dominant feature of global political agendas. In the mid-1980s, 'green' political parties and NGOs in North America and Western Europe appeared highly influential in shaping political ideas, and mainstream political parties grappled with new environmental practices and vocabularies. Ecological disasters such as the 1986 Chernobyl nuclear explosion helped to create a situation where environmental issues enjoyed an unprecedented profile. In the 1990s, especially following the 1992 Rio summit, green issues appeared to lose political and electoral impact. Public attention in the United Kingdom has occasionally been dominated by high-profile environmental cases, such as the dumping of the Brent Spar oil platform in 1995. Greenpeace were able to prevent the platform being dumped at sea, but their shock tactics were later heavily criticized by the government and independent scientific observers for misrepresenting the inherent dangers of deep-sea dumping. Public support for many of Greenpeace's actions has not translated to support for the Green Party at UK general elections.

Further reading

See G. Piel's book *Only One World* (United Nations, 1992), J. Palmer's chapter 'Towards a sustainable future' in *The Environment in Question* edited by D. Cooper and J. Palmer (Routledge, 1992). Also see A. Dobson's *Green Political Thought* (Routledge, 1990), the volume edited by G. Prins *Threats Without Enemies* (Earthscan, 1993), L. Elliot's *The Global Politics of the Environment* (Routledge, 1998) and J. Gupta's *The Climate Change Convention and Developing Countries: From Conflict to Consensus?* (Kluwer, 1997). G. Graham has written *Ethics and International Relations* (Blackwell, 1996) and S. Breslin has published 'China's environmental crisis in a global context', *Global Society* **10**: 125–44. For critiques of environmental security see S. Dalby, 'Security, modernity and ecology: the dilemmas of post-Cold War security discourse', *Alternatives* **17**: 95–134; P. Stoett, 'The environment enlightenment: security analysis meets ecology', *Coexistence* **31**: 127–47; and J. Seager's *Earth Follies: Coming to Feminist Terms with the Global Environmental Crisis* (Routledge, 1993).

Chapter 7

The globalization of humanitarianism

The Universal Declaration of Human Rights, adopted by the United Nations General Assembly in 1948, is a global charter of the rights and responsibilities of individuals and states. If properly implemented, the UN charter would, in principle, undermine a fundamental assumption of international politics (in realist and liberal thought): non-interference in the domestic affairs of particular countries, i.e. the principle of state sovereignty. There is therefore a potential tension between the rights and obligations derived from nation states and the human and legal rights endowed by the international community (Mullerson 1996).

At first glance, the geographical and legal implementation of human rights should provide an excellent case study for geopolitics and global politics. Unfortunately, the international community's willingness to protect rights over the last fifty years has been patchy, as the human rights trials currently in progress (Bosnia, Rwanda, Burundi and Nazi legacies) would seem to attest (Mullerson 1996; Baylis and Smith 1997). In the last fifty years, the geography of human rights protection has tended to favour the wealthier Northern states and their citizens rather than the poorer Southern states, some of which have experienced brutal regimes and massive human rights violations (Chapter 3).

Government pledges by the UK foreign secretary, Robin Cook, to promote a new human rights agenda have been derailed by the admission in August 1997 that defence agreements with the Indonesian government would continue in spite of opposition from human rights campaigners in and outside East Timor. The UK-based Campaign Against the Arms Trade (CAAT) and Ann Clwyd, the chairperson of the House of Commons all-party group on human rights, had called on the foreign secretary to ban all arms exports to Indonesia, but for all the rhetoric, states such as the United Kingdom often seem unwilling to place human rights at the top of the diplomatic agenda in the face of commercial and security interests. In a more recent case in October 1998, the UK foreign secretary faced further embarrassment when the British courts ruled that the former dictator of Chile, General Augusto Pinochet (alleged to have overseen the murder of three thousand Chilean and other citizens between 1973 and 1977) did not enjoy diplomatic immunity in the United Kingdom. Human rights groups in Chile, Spain and the United Kingdom have argued that this former head of state should face charges

Figure 7.1 A UN assistance mission in the Great Lakes region of Central Africa; the mission is located at Ndosha Camp in Zaire, now the Democratic Republic of Congo. (Photo by J. Isaac; reproduced with permission of the United Nations)

under international law for human rights abuses. The case of General Pinochet remained unresolved in early 1999.

This chapter concentrates on the position of human rights within the global political agendas of the post-Cold War era due to the growing international profile of human rights and the increasing demands of humanitarian assistance (Figure 7.1). The first part of this investigation will consider some of the serious political problems relating to the conceptualization and defence of human rights in the absence of universal consensus in favour of human rights *per se*. For post-colonial critics, human rights are considered to be part and parcel of a Western doctrine of rights, which is insufficiently sympathetic to the diverse cultures and communities of the world. From that vantage point, human rights can only be culturally specific rather than universally applied. The second part of this chapter considers humanitarian intervention in world politics, and it questions whether any intervention can be justified on the basis of human rights (Vincent 1974). Should poverty and underdevelopment be grounds for non-violent interventions, which some countries and observers demanded for places suffering from genocide, ethnic cleansing and famine? It could be argued that the humanitarian needs of many citizens have been seriously neglected considering that around two million children have been killed in the present decade and that over two billion people lack clean and regular drinking water. The final section returns to the problem of incorporating these issues into political agendas dominated by states, international organizations and national interests.

Conceptualizing human rights: definition and implementation

Human rights have long occupied a place in Western political thought, from the thirteenth-century Magna Carta to the eighteenth-century Bill of Rights. This sustained interest in rights, responsibilities and natural law encouraged humanitarian organizations such as the Anti-Slavery Society in the nineteenth century to extend Western conceptions of human rights to non-European peoples. While the spread of humanitarianism is frequently described as a Western phenomenon, it is none the less important to recognize that non-Western societies and faiths have demonstrated considerable compassion and responsibility towards the vulnerable, weak and endangered. For the purpose of this chapter, however, attention will concentrate on international law and practice which has evolved since 1945.

The definition of human rights has been a contested affair. For Western observers, political and legal rights, such as the freedom of assembly, have been conceived in a liberal democratic tradition which stressed the rights of citizens in relation to the state, although the relative values often varied depending on the interaction of liberal and democratic agendas (Gearty and Tomkins 1996). The extension of rights to women and ethnic minorities was a long-drawn-out process even in these liberal democratic nations. For socialist observers, human rights have been conceptualized in broader terms to include social and economic rights, such as full employment. The creation of the Soviet Union in 1917 was premised on the belief that social and economic rights would compensate for the loss of formal political rights (Lane 1996). The International Covenant on Economic, Social and Cultural Rights in 1966 recognized that there were certain social and economic rights, and it effectively confirmed that the UN Commission of Human Rights had failed to agree on a universal codification of the declaration. This 1966 covenant came into force in 1976 at the same time as the International Covenant on Civil and Political Rights.

Post-colonial critics and indigenous groups have argued that these definitions of human rights (whether civil, political or economic) ignore the fact that some groups are more concerned to preserve their culture and environments. The most common meaning attached to the term 'post-colonial' refers to the ending of predominantly European colonialism and the emergence of post-colonial states such as the former British colonies in Africa and Asia. In this context, it refers to the removal of external forms of control and exploitation as witnessed during the era of the British Empire. While it is often argued that new forms of colonialism may exist in the form of economic, diplomatic and political manipulation, most accept that the formal European and American empires of the nineteenth and twenthieth century have virtually disappeared. The term 'post-colonial' also applies to ways of thinking about the world, and so-called post-colonial critics have argued that many conceptions of politics, human rights and economic management are based on Western assumptions of individual freedom, liberal democracy and market economies. African and Asian observers believe there can be few universal human rights because this would imply that all the world agrees on what actually constitutes human rights (R. Walker 1988).

The 1981 African Charter on Human and Peoples' Rights recognized that the demands of African peoples were different from other peoples and implicitly argued that human rights have to refer to more than formal political and legal rights. The right to tribal survival and the preservation of African cultural values was a central theme in this document, unlike the UN Universal Declaration, which focused on the rights of peoples rather than individuals. In a similar vein, the American Convention on Human Rights (1969) recognized specific Latin American concerns over human rights protection. Accordingly, universal human rights are perceived by many African, Latin American, Asian and Pacific observers to be little more than an imposition of Western values and belief systems.

Feminist writers have argued that Western understandings of human rights are also rooted in patriarchal assumptions of the role of women and families (Pettman 1996: 208–11). During the United Nations Decade for Women (1975–1985) it became readily apparent that women were systematically disadvantaged in terms of sexual rights, property ownership, legal protection and through access to health and education (Enloe 1989; Afshar 1998). In that respect, the existing norms and values reflect the experiences of men, hence they cannot be universal because they ignore the experiences of women. Moreover, the dominant conceptions of human rights fail to recognize that women are more often in need of protection in the home, where most of the violence occurs, rather than in the public sphere. This argument could also be extended to children, the disabled and the elderly, but the human rights discourse tends to focus on formal politics and the public sphere of societies.

This exclusion of gender from the universalistic conceptions of human rights is further compounded by a lack of recognition regarding social and economic rights. The distribution of global income is highly unequal, with only an estimated one per cent of property, land and financial resources being held by women (cited in Bretherton 1996: 256). A commitment to social and economic rights would have to be grounded in an appreciation of the widespread exclusion of women from the ownership of wealth. Changing the rights of women would therefore involve some fairly fundamental reorganization in the world's political economy. It was not until 1984 that the United Nations Commission on Human Rights recognized how domestic violence against women should be a subject for human rights discussion. Subsequent UN conferences in Vienna (1993) and Beijing (1995) have continued the political debate and policy discussions over the role of women in international politics and humanitarianism (Haynes 1996).

Defining and then defending human rights remains a problematic venture. Some liberals argue that the right to cultural survival and/or environmental security are not really 'rights' in the first place. Alternatively, there has been much criticism from organizations such as Amnesty International (see Figure 7.2) that well-established political rights such as those governing torture and illegal imprisonment are frequently neglected in favour of ensuring compliance.

Figure 7.2 Amnesty International poster.

Human rights lawyers and critical political commentators endorse the notion that legal obligations regarding the defence of human rights often appear to be sacrificed in the realpolitik of national interests. From a feminist perspective, the protection of women's rights also tends to be haphazard, as witnessed in Bosnia and Rwanda, where the mass rape and mutilation of women and female children was endemic. In 1994 a special rapporteur on violence against women was finally appointed to highlight the inadequacies of current protective measures based on the civil and/or public sphere.

Enforcing human rights: legal and geographical variations

The UN Declaration of Human Rights (1948) remains the landmark document in terms of international legal obligations. By building upon the sentiments of the UN charter and the establishment of the UN Commission on Human Rights, it embodied a series of so-called first-generation rights to political freedoms, such as the right to freedom of speech and choice of religious denomination. It was adopted in the UN General Assembly by 48 votes to nil, with eight abstentions (Soviet Union, Ukraine, Byelorussia, Czechoslovakia, Poland, Yugoslavia, Saudi Arabia, South Africa). This Declaration was the first attempt to define a code of international behaviour and to stress that universal standards should be respected by member states of the United Nations. It was composed of thirty articles, which covered civil and political rights as well as a range of economic and social rights. Article I enshrined the principle that 'all men are born free and equal in dignity and rights'. Other articles of the declaration dealt with the freedom to choose a religion, the right to education, the right to be secure from the threat of torture and illegal imprisonment.

To most Western observers, the UN declaration was not a problematic document, as it secured political and legal rights already enjoyed by the majority of these nations. While there was some debate to establish whether the declaration should seek to protect a narrow range of rights or to be more progressive, most Western nations were in accord with the importance of protecting civil and political rights. The declaration was far more problematic for the South African apartheid regime, which abstained from the vote because it refused to accept that its domestic authority would be challenged by these human rights obligations. The apartheid system, established in 1948, had effectively excluded the majority of the population from the political system (Chapter 3). The Soviet Union and its political allies also abstained, because it was argued that the declaration took no account of social and economic rights. The Saudi Arabian monarchy objected because the freedom to choose one's religion (article 18) violated Saudi law, which outlawed religious denominations other than Islam. Coming from one of the few non-Western nations, Saudi Arabia's objections to the universalism of the declaration were to be reinforced at a later stage by the decolonized nations of Africa and Asia in the 1950s and 1960s. Interestingly, Saudi Arabia has never ratified other international covenants such as those concerning civil and political rights (1966).

It was a remarkable feat that the UN Declaration of Human Rights was negotiated at the start of the Cold War. For the Soviet Union, the declaration was a profoundly Cold War document designed to attack the Soviets and their allies for totalitarian governance. Joseph Stalin argued that the United States was represented as the model for other nations to follow in terms of governance and the provision of human rights, but that alternative conceptions of democracy and human rights were effectively marginalized by the underlying impulse of this declaration. Unsurprisingly, Soviet leaders tended to raise other issues, such as the racial marginalization of Black people in the United States (and their subsequent struggle for civil rights there); and the United States condemned Russia for violations of civil and political rights.

This underlying geopolitical context prevented the introduction of a more binding covenant and restricted the geographical and legal enforcement of human rights obligations to be carried out by the international community in a uniform and even-handed manner. In many parts of the world such as Latin America, Southern Africa and Southeast Asia, the United States and its allies were prepared to overlook massive human rights violations by pro-Western military regimes in order to ensure that communism would not flourish in these regions.

In the 1970s Argentina was governed by a series of brutal military regimes which launched a massive and violent campaign against so-called subversives under the label of the Dirty War. The military juntas argued that Argentine national security was being compromised by left-wing revolutionary elements in society, whose sporadic guerilla activities in certain parts of the republic were cited as an excuse for a violent national security strategy (Nino 1995). Individuals connected to trade unions, professions, the Catholic church, the media and universities were targeted for persecution, and Amnesty International estimated that over ten thousand people were executed, tortured and/or simply 'disappeared' during the period between 1976 and 1981. Victims were often dragged off the streets and bundled into cars, which then travelled to a network of detention and torture centres. Afterwards many of the bodies were thrown out of armed forces' planes into the shark-infested South Atlantic.

In 1977 the Las Madres de la Plaza de Mayo was formed by mothers of the missing relatives, who gathered in demonstration in the central squares in Buenos Aires. Employing the tactics of peaceful resistance, the group's unwavering vigil against the unmitigated brutality eventually forced the military regime to confront the violence of the Dirty War. It was hoped that the fate of the many missing victims would be resolved. To this day, every Thurday afternoon a group of mothers still gathers at the Plaza de Mayo in Buenos Aires, holding aloft photographs of the 'missing'.

For citizens in East Timor, Argentina, Chile and South Africa, 'universal' human rights were often a mirage. During this era the United Nations undertook some limited human rights and peacekeeping work in Cyprus, Korea and the Middle East, but Soviet violations of human rights in Hungary and Czechoslovakia, although condemned, were not actively challenged, because the dangerous machinations of the Cold War frequently prevented powerful Western nations from intervening unless they occurred in areas of the world considered to be strategically unimportant and/or governed by weak states.

Increasingly, especially during the aftermath of the Cold War, the most vigorous defence of human rights was often stimulated by human rights organizations exposing violations on television and in print. In the middle of the Cold War, the pressure group Amnesty International was founded in the

United Kingdom to monitor human rights abuses and violations around the world. From 1961 onwards the London-based organization has campaigned on behalf of those illegally imprisoned, tortured and/or denied basic human justice. Amnesty International relies on voluntary donations and private subscriptions to fund campaigns on specific cases such as the imprisonment of Nelson Mandela in South Africa, the enduring Indonesian violence in East Timor and the human rights abuses in Myanmar, China and Argentina. Their annual evaluation of human rights has also included criticism of governments; the UK government was criticized for repressive security policies in Northern Ireland when the police and armed forces denied basic civil and legal rights to 'terrorist suspects'. Amnesty International remains one of the largest groups committed to the cause of human rights; and states such as the United Kingdom, the United States and France have been goaded into taking action against human rights violators by the efforts of such pressure groups.

While there have been many examples of Western governments refusing to intervene on behalf of oppressed peoples for commercial and geopolitical reasons, it has to be acknowledged that enduring problems exist regarding the defence of universal human rights. Universalism – in any form – can be insufficiently attuned to regional and local differences around the world. The Declaration of the Principles of Indigenous Peoples (1989), for instance, recognized that human rights had to include the protection of the customs and practices of indigenous peoples. Some liberal observers maintain that universal human rights can also lead to the destruction of cultural differences.

Can human rights ever be universal?

It should already be apparent that the idea of universal human rights is conceptually controversial and politically problematic. For supporters of universality, human rights are derived on the basis of a moral argument regarding the intrinsic and equal worth of human beings. Article 1 of the UN declaration reflects this particular philosophical position by acknowledging that 'all human beings are born free and equal in dignity and rights'. The underlying concept of universal rights has been challenged by Western critics who take issue with the assumption that human nature is based on the capacity for moral reasoning and rational action; post-colonial critics have questioned the Western philosophical assumptions of individual rights; and feminist critics have censured the gendered assumptions of human rights discourse and practice.

On a more optimistic note, some observers, such as Francis Fukuyama (Chapter 1), believe that with the emergence of liberal democratic governments in the 1980s and 1990s in Latin America, Eastern Europe, Africa and Asia, the spread of democracy enables a greater number of countries to share a particular moral and political consensus. Many people in Central and Eastern Europe greatly value their new-found civil and political rights, but the problem with this kind of argument is it ignores the fact that regions such as the Middle East have shown little inclination towards democracy and/or neglects the fact that many Asian nations have consistently rejected (Western) civil and political conceptions of human rights. The 1993 World Conference on Human Rights witnessed Asian, particularly Chinese, opposition to universal concepts of rights.

One way forward for the human rights agenda might be to concentrate on identifying human wrongs rather than rights. It has been proposed that international law should seek to develop a new code of human wrongs, which could then be used to delegitimize certain actions. Supporters of this strategy have pointed to the 1948 Genocide Convention to illustrate the successful use of international law for the prosecution of officials and soldiers responsible for the Rwandan massacres in 1994–1995. This convention, based on the experiences of the 1945–1946 Nuremberg war crimes trials, introduced into international law the concept of crime against humanity. The prosecution of German military and civilian leaders relied on the assumption that universal and inalienable rights were grossly violated during the Second World War, thus a new legal precedent had been created in 1945–1946. However, ensuring this compliance has been problematic, as testified by the massacre of over two million people in places such as East Timor, Cambodia and Rwanda. It is also proving difficult for the international community to prosecute those suspected of massive human rights abuses (Nino 1995).

The alternative proposed by some philosophers seeks the promotion of non-foundational human rights to overcome the universal impasse. Grounded in an appreciation that cultural relativism can and has been used by repressive regimes to justify massive human rights violations, this approach would seek to match cultural traditions with acceptable forms of human rights. At the 1993 UN Conference on Human Rights, India and China condemned universal human rights for being little more than an extension of European understandings of rights and human freedom. The challenge for Western nations determined to uphold a tradition of universal human rights would be to demonstrate that political and social life is preferable in a context where human rights are respected as opposed to violated. International standards on political rights regarding issues such as torture and genocide would have to be protected in tandem with a commitment to respect particular cultural variations. No Chinese leader has ever claimed the *de jure* right to torture political suspects and/or imprison dissidents without trial.

In the post-Cold War era, the protection of human rights remains precarious. Upholding universal human rights is unlikely to be achieved without some consideration of the material conditions of life. Meanings of needs, justice and ownership within different societies would have to be recognized within discourses on rights and democracy (Baylis and Smith 1997). When the UK foreign secretary met the prime minister of Malaysia in August 1997, newspaper reporting of the meeting indicated that Malaysian Premier Dr Mahathir Mohammed was unhappy with Robin Cook's rigid adherence to the UN declaration in their discussion of the human rights situation in Southeast Asia. Dr Mahathir argued that the declaration was scripted by and for the rich, hence it failed to address Asian and African values. Malaysia later expressed reservations towards UK proposals for further sanctions against Nigeria (in the wake of the Ken Saro Wiwa hanging in 1995) as presented to the Commonwealth summit in October 1997. Dr Mahathir has declared time and again that he does not believe democracy to be any better than dictatorship, and that sanctions directed against dictatorial regimes because of human rights abuses merely tend to hurt the poor of that country.

In the late 1990s, therefore, no global consensus on human rights exists in spite of reported human rights abuses attributable to 150 countries. Although human rights monitoring has improved through the appointment of a high commissioner for human rights (1994), there remains much to do in terms of ensuring basic human rights compliance. Other UN-based organizations, such as the Committee on the Discrimination against Women (CEDAW), have attempted to address the problem of women's rights, but underfunding has hampered progress in recent years. Reconceptualizing human rights is only part of the problem, and future progress will depend on the development of an effective machinery designed to protect and monitor human rights.

Humanitarian intervention in the post-Cold War era

Television coverage of human suffering has the capacity to influence or even determine government decision making. This realization has been one of the most talked about features of the post-Cold War era. Media circles call it the CNN factor, in the wake of several humanitarian missions in Europe, the Middle East and Africa. For some observers, the power of television to influence government decision making is perceived as a positive feature of global politics because it can mobilize the international community into taking action in defence of massive human rights violations. Yet political leaders and their advisors have cautioned that television coverage can be selective in the sense that the media concentrates on some places at the expense of others such as Kashmir and Angola. Newspaper reporting can often neglect Third World human rights issues in favour of 'European' events, such as the recent Russian political crisis and the peace process in Northern Ireland (Figure 7.3). Finally, some organizations responsible for humanitarian aid, such as the International Committee of the Red Cross, have complained that media coverage of human suffering can be too simplistic and it often unwittingly conceals the global processes that help to perpetuate poverty, hunger and human rights violations. The media reporting of the Rwandan massacre, from April 1994 onwards, concentrated on ethnic and tribal divisions within Rwanda but failed to explore the role played by great powers such as France in supporting a brutal and murderous regime for over twenty years (Prunier 1995).

Humanitarian intervention has therefore emerged as one of the leading areas for debate in discussions on global politics and human rights (Agnew 1992; O'Loughlin 1992). Discussion on the role and purpose of humanitarian assistance has extended the limited realist view of political life, which is sceptical of international cooperation and/or critical of global aspirations. The provision of relief for distressed peoples and the protection of basic human rights have emerged as central themes in these new humanitarian debates (Pugh 1996; Weiss and Collins 1996). Former UN secretary general Dr Boutros Boutros-Ghali (Figure 7.4) has published *An Agenda for Peace*; it places

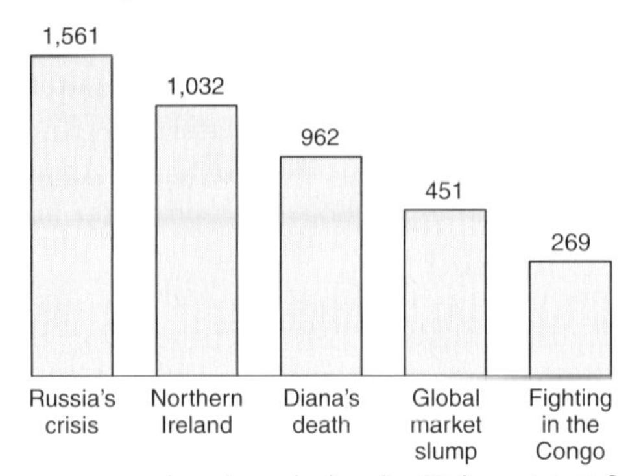

Figure 7.3 Newspaper coverage in column inches for 29 August to 3 September 1998. (Data obtained from the *Guardian*, *The Times*, *Daily Telegraph*, *Independent*, *Sunday Times*, *Observer* and *Daily Mail*)

Figure 7.4 Boutros Boutros-Ghali, sixth secretary general of the United Nations. (Photo by M. Grant; reproduced with permission of the United Nations)

human rights (HR) and peacekeeping at the top of the UN's agenda and demands that member states should commit themselves to the funding of HR-related projects (Boutros-Ghali 1992). Boutros-Ghali was also a firm supporter of Chapter VII of the UN Charter in order to justify outside interference in the affairs of a state which is guilty of massive HR abuses. In April 1991 he supported resolution 688, which authorized Western jets to protect the airspace above northern Iraq when fears surfaced that the Kurdish people were vulnerable to attack from Saddam Hussein's airforce.

At the heart of these examples lies a geopolitical paradox: How can humanitarian intervention be justified when it occurs within an international political system premised on state sovereignty and norms of non-intervention? For the last three hundred years, a Westphalian society of nations has been based on the assumption that states were responsible for their own affairs and that, by implication, other states could not intervene in the domestic affairs of their neighbours. In the aftermath of the Holocaust and the creation of the United Nations, a new wave of human rights charters and conventions changed the relationship between nation states and international society (A. Smith 1995). Intervention could be justified in terms of either self-defence or in an attempt to prevent murderous states committing massive human rights violations. Since the ending of the Cold War, these ideas have been expanded to include consideration of non-military and military forms of intervention by states and non-states; but because of the concerns expressed by Asian and African states that humanitarian intervention (as with human rights) could be used as an excuse for further involvement by the great powers in the affairs of the weaker nations, these discussions have been confined to the West.

Humanitarian intervention: for and against

The dilemmas concerning humanitarian intervention (HI) touch upon some of the critical issues facing contemporary world politics. The apparent moral imperative to relieve the suffering of others, has been much debated within academic disciplines as to the legality of intervention. The Commission on Global Governance has argued that intervention is justified in cases where massive violations of human rights have occurred, and it believes an international response would be required to prevent a further loss of life. The justification of outside intervention and the role of force in order to secure humanitarian objectives have been vexing problems (Table 7.1). Some of the issues regarding the use of military force for the purpose of humanitarian intervention are addressed using examples from recent UN involvement in Somalia (1992–1993), Rwanda (1994–1995) and Bosnia (1993–1995).

The use of military force to support HI has been called into question after the televised failures of the Somalian and Rwandan humanitarian operations. The Americans were widely seen to be the principal supporters of this intervention, but US troops despatched to the Horn of Africa in December 1992 quickly became engrossed in street-to-street fighting with local factions and several so-called warlords. By October 1993 President Clinton ordered the

Table 7.1 Complex humanitarian emergencies

Country	Population at risk[a]	Includes[b]	Political environment
Angola	3.7 million	3 million refugees in Congo, Zambia and Zaire	Civil war Intensified hostilities Limited government ability to support relief operations
Bosnia and Hercegovina	2.5 million	1.1 million refugees primarily in Croatia, Germany and Serbia	Little government control All factions periodically oppose relief to other groups
Rwanda	4 million	2 million refugees primarily in Burundi, Tanzania and Zaire	Ethnic warfare
Sudan	3 million	0.4 million refugees primarily in Ethiopia, Uganda and Zaire	Ongoing insurgency All sides use relief as weapon Government opposes relief to south and to non-Muslims in north
Zaire	0.6 million	75,000 refugees primarily in Burundi, Tanzania and Zambia	Government cannot assist relief Little or no civil authority Crime and extortion

[a] The term 'population at risk' indicates those people who are in need of, or who depend upon international aid to avoid large-scale malnutrition and deaths, including refugees, internally displaced persons, and others in need.
[b] These refugees are included in the figures under population at risk.
Source: Adapted from Weiss and Collins (1996).

retreat of the US military from Somalia after the death of eighteen soldiers in an armed confrontation with a local militia. Within the United States, television pictures of dead US servicemen being dragged through the streets of Mogadishu prompted much internal debate over the effectiveness of intervention, and there were calls for a reappraisal of the UN relief operation. Notwithstanding the tragedy of these and other deaths (such as the hundreds of Somalis who were killed during US military operations), there are strong reasons for a critical examination of the intervention:

- Article 2 of the UN charter enshrines the principle of non-intervention. This means that states must respect the sovereignty of other states and refrain from intervention in their affairs. HI is an act which seeks to intervene in the domestic affairs of another state. At the very least, military-based HI would be illegal under the UN charter unless one could demonstrate that wider human security was at risk. In 1978 Tanzania invaded Uganda with the purpose of overthrowing the murderous regime of Idi Amin. During a period of eight years over 300,000 people were murdered. Tanzania argued that these activities threatened the security of her own people because the Ugandan security forces had crossed the border in search of further victims. This intervention, while not expressed in humanitarian outcomes, had a humanitarian benefit in terms of removing Amin from power.

- It has been argued that states sometimes act for ulterior motives. It has been alleged that France proposed HI in Rwanda during the massacre of Tutsis in order to maintain her influence in Francophone Africa rather than to relieve the human suffering. There is no question that HI would have had the effect of raising France's profile in a region where American influence was growing at the expense of the French. Critics maintain that many large states are not committed to HI; the recent American budget for the development agency USAID has been reduced and US foreign aid donations have dropped to 0.15 per cent of GDP.
- The principles of HI have and will continue to be inconsistently applied by the international community. Sceptics point to the contrasting experiences of the Bosnian Muslims and the Iraqi Kurds. The Kurds were judged to be needy of HI, and in 1991 the American-led operation established a safe-haven policy in northern Iraq for the purpose of protecting the Kurds and the so-called Marsh Arabs in the south from Saddam Hussein's regime. In contrast, the international community was unprepared to intervene actively in the defence of the Bosnian Muslims in 1992–1994 (Gow 1997).
- Definitions of humanitarian acts, such as seeking to prevent the suffering of others, can be problematic in the sense that our understandings of humanitarianism can change over time and alter over space. Public attitudes towards slavery in Britain have changed from benign acceptance to outright rejection over the last two hundred years. Understandings of human suffering (as with human rights) can also be culturally specific.
- Defining common principles of HI will always be difficult because they rely on the international community's agreement to place individual and communal justice above the principle of sovereignty and non-intervention. In the absence of any common principles, it might be better to avoid legitimizing HI within the canons of international law (Mullerson 1996; Pugh 1996).

For sceptics of HI there are good reasons to resist the legalization of such practices. In spite of the apparent moral imperative to relieve human suffering, there is ample evidence doubting the wisdom of HI.

In an alternative vein, a considerable legal and moral case can be brought to bear in favour of HI. The arguments are as follows:

- The UN charter commits states to protecting fundamental human rights and to prevent abuses. In that context, HI could be seen as a legitimate mechanism for the preservation and protection of human rights. Recent UN resolutions such as 43/101 and 45/100 have enshrined the right of NGOs to assist in the provision of aid to troubled areas and demanded that so-called corridors of tranquility should be created in order to assist the

Figure 7.5 UN Protection Force (UNPROFOR): a Canadian soldier visiting two young boys of Croatian descent. (Photo by S. Whitehouse; reproduced with permission of the United Nations)

delivery of aid. NGOs such as the International Committee of the Red Cross have emerged as important sites of advice and consultation.

- In a moral setting, it has been argued that the international community has an obligation to ensure that states respect basic human rights. Under the canon of international law dealing with genocide and human rights, HI would be a morally and legally legitimate form of intervention if there were evidence of massive oppression and violation.
- There is evidence to suggest that HI is part of customary international law and that states have recognized in the past that intervention might be justifiable. The Vietnam invasion of Cambodia in 1979, although not enframed in humanitarian terms, demonstrated that the international community will tolerate certain actions if they are seen to produce humanitarian benefits like the ending of a genocide by murderous figures such as the late Pol Pot.
- With the termination of the Cold War, there is a real possibility for the international community to develop a new moral and legal consensus on norms of behaviour. The UN's budget for peacekeeping and humanitarian projects has grown substantially, and in 1994–1995 the UN was involved in seventeen operations with the assistance of 75,000 personnel (Figures 7.5 and 7.6).

In terms of clarifying the legal and moral position of HI, the way forward is contested. For supporters of HI, the international community should make

Figure 7.6 UN Protection Force (UNPROFOR): a Canadian civilian police (CIVPOL) officer (third from the right) talking to Croatian police officers in Daruvar. (Photo by S. Whitehouse; reproduced with permission of the United Nations)

every effort to develop a consensus around which the principles of HI could be firmly elucidated. For much of the Cold War, HI was rarely considered by the international community because strategic rather than humanitarian intervention tended to dominate international affairs. The 1990s presented new opportunities to consider how humanitarian outcomes could be secured through intervention. But sceptics claim that if HI was ever formally legitimated by the international community, it could undermine the contemporary international order based on the principles of sovereignty and non-intervention. Realists suggest that the international community should not seek to develop new norms of governing HI because they might undermine the existing international system.

Contemporary humanitarian intervention: enduring tensions

In the contemporary era, Western sceptics of HI have in some sense been overtaken by events. Televised reporting of human suffering has had a powerful influence on domestic public opinion in regions such as Europe and North America. For better or worse, political leaders are having to confront HI's pros and cons in an unprecedented fashion. Television coverage can be used to force political leaders to think about, even commit to, intervention for the relief of suffering. Alternatively, media reporting of humanitarian disasters can also make HI politically unattractive in places such as the United States. For non-Western nations, however, the current propensity for HI discussions in the West is regarded with some suspicion. China has been consistently sceptical of Western motivations for intervention, and as a veto-carrying member of the UN Security Council, it is likely to be hostile to granting HI a legitimate status.

An individual operation's chances for success or failure have to be established, because this will have direct implications for the future legitimacy of HI in general. The use of force is considered in the deployment of HI because in the recent past it has been necessary to protect the operations of the UN agencies. The role of NGOs, such as the Red Cross and Médecins sans Frontières, should also be considered because it is clear that states (contrary to realist assumptions) have to coexist with other non-state parties in the international humanitarian arena.

It has been noted that HI in the post-Cold War era has rarely been legitimated by the UN Security Council on the basis of humanitarian grounds alone. While humanitarian considerations were undoubtedly important in motivating UN operations in Somalia and Kurdistan, Western political leaders frequently drew upon other factors to justify the relief of massive human suffering. In the case of Kurdistan, UN resolutions (e.g. resolution 688) governing military-backed humanitarian intervention were justified by Western leaders on the basis to 'protect' the international peace and security of the region in the light of Saddam Hussein's repression in northern Iraq. Other observers, such as China, objected to the approval of military enforcement action (such as the declaration of a no-fly zone in northern Iraq) because they were concerned this action would establish a precedent for use of military force in the defence of human rights. The refusal to endorse a military enforcement mandate by China and the Soviet Union meant the United States and her allies used resolution 688 to employ parts of the UN charter to justify military intervention.

In contrast, resolution 794 was cited to defend the American-led relief operations in Somalia to alleviate the suffering. Military intervention was justified and approved by the Security Council because the Somali state had collapsed and violent civil war had broken out. Moreover, in contrast to the Kurdistan operation, there was no direct opposition to the intervention from a *de facto* leader of state. The norms of non-intervention and state sovereignty had not been undermined by this US-led intervention. As a failed state, Somalia was not even considered by sceptics such as China to present a general precedent for HI, and the final draft of resolution 794 contained numerous references to the unique and unusual nature of the Somali case.

The effectiveness of the UN Somalia operation was modest (United Nations 1996). In the short term there can be no doubt that the operation led by the United States did relieve the suffering of starving civilians. In the longer term, however, the plan to demilitarize Somali society was called into question because the United Nations and the United States apparent neutrality over humanitarian aid appeared to be replaced by a desire to capture particular warlords for the purpose of imposing order. Television coverage of US gunships hovering over the streets of Mogadishu appeared to confirm this humanitarian operation had been transformed into a military exercise.

The UN response to the crisis in the former Yugoslavia (1992–1995) also highlighted the complex interrelationship between moral claims regarding human rights and norms regarding state sovereignty and non-intervention (Woodward 1995; Weiss and Collins 1996). This example of humanitarian intervention had substantial consequences for post-Cold War global and European geopolitics. The crisis in the Balkans was the first example of the United Nations operating in the heart of Europe as opposed to the so-called Third World, and billions of dollars were invested in conflict and refugee management. However, the apparent failure of the United Nations to intervene decisively in 1992–1993, in the face of human rights violations and ethnic cleansing, caused a major crisis of confidence in HI provision. The United Nations did not attempt to stop these violations; instead it approved a form of peacekeeping that attempted to combine humanitarian assistance without active engagement with the parties responsible for genocide and the destruction of multi-ethnic communities. Former UN commanders such as General Francis Briquemont frequently complained that UN Security Council resolutions regarding the former Yugoslavia were not being applied in the field because of a lack of political will to enforce peace accords. The UN peacekeeping force in Croatia was only given a US$250 million budget when experts had estimated that around US$700 million was required to fulfil the UN peacekeeping and humanitarian objectives (Weiss and Collins 1996).

The deployment of fourteen thousand UN peacekeepers in 1992 to the Republic of Croatia (Figure 7.7) was not able to prevent the spread of war in the former Yugoslavia, or to empower Croatian refugees to return to former villages and towns. The imposition of no-fly zones in Bosnia in 1993–1994 was rarely enforced, and the so-called protection of safe havens was piecemeal and haphazard. Over forty Security Council resolutions urging the parties to call an end to the fighting in the former Yugoslavia were passed with little impact (Mullerson 1996). HI was justified in 1992–1994 on the basis of threats to international peace and security, but was largely ineffective because relief convoys were not adequately protected from Serbian and Croatian armed factions. By late 1993 eighty thousand UN peacekeepers had been deployed in the Balkans for the purpose of humanitarian assistance and conflict resolution, and in spite of the massive increase in peacekeepers, the debt-ridden United Nations proved incapable of enforcing peace in the region.

In February 1993 it was mooted that calls for HI should be ended by the United Nations because of the persistent failure of warring parties to accept agreements over access to aid for civilians. Some commentators were already calling into question the financial and political wisdom of HI in the face of other resource claims regarding poverty and preventable diseases. Others demanded the United Nations create a new humanitarian delivery unit with clear rules and procedures for the delivery of HI (Weiss and Collins 1996). Inadequate military and humanitarian action in the former Yugoslavia was compounded by doubts among Western allies about the effectiveness of air strikes, sanctions and on-the-ground military intervention. Media reporting of human suffering combined with the 'something must be done' factor cast considerable doubt on the effectiveness of HI.

Figure 7.7 Before the destruction of Yugoslavia, the towns of Zagreb (top) and Dubrovnik in Croatia had been popular holiday destinations for Western Europeans during the Cold War. Tourism collapsed after 1991 as the Yugoslav Republic descended into violent conflict. (*Photos*: Klaus Dodds)

Within the United States, considerable soul-searching ensued regarding the provision of HI and the use of military force in the wake of the Yugoslav crisis. The experience of Somalia led conservative commentators such as Senator Sam Nunn to argue that the United States should not commit troops to Bosnia for

Figure 7.8 The geopolitical condition of Bosnia after the 1995 Dayton Peace Accord, which split the republic into two parts: a Serb-controlled Republika Srpska and a Bosnian Federation. Sarajevo, the capital city, was placed under UN supervision. Some of the worst atrocities of the Yugoslav conflict (1992–1995) occurred in towns such as Banja Luka, Gorazde, Gornji Vakuf, Mostar, Srebrenica and Tuzla.

the purpose of conflict resolution. The 1995 Dayton Accord, which finally secured a modicum of peace and territorial stability in the region, actually confirmed the ethnic cleansing gains of the powerful Serbian and Croatian factions. Fifty thousand NATO troops, deployed in a peacekeeping role, effectively ratified the status quo ante. So in terms of humanitarian outcomes (as opposed to motives), HI in the former Yugoslavia achieved mixed results. Although it brought some relief to the suffering of civilians in the war zones, it failed to prevent the killing of Bosnian Muslims by Serbian forces in towns such as Srebrenica (Honig and Both 1996). As to the future, time will tell whether the reconstruction and development of the region has been facilitated by HI (Figure 7.8). The toll of the Yugoslav wars was, however, dreadful in the sense that an estimated 200,000 died, 2.7 million people were made homeless and six million anti-personnel landmines remain buried in the region (Figure 7.9).

Accordingly, China and India have argued that state sovereignty does not permit outside intervention, even though powerful states such as the United

Figure 7.9 Bosnia: typical destruction left by the war. Selim Ornanovic, a Bosnian muslim refugee in London for four years, returns to his destroyed home in a village near Kluj. The village was occupied by Bosnian Serbs before being retaken by the Bosnian army. Selim spent six months in a Serbian concentration camp. (*Photo*: Howard J. Davies/Panos Pictures. © Howard J. Davies 1996)

States and the United Kingdom use the UN charter's definition of 'international peace and security' to suit their own agendas. Selective application of this definition means that countries like Somalia are deemed worthy of HI whereas neighbours such as war-torn Sudan are ignored. Some humanitarian workers have reached similiar conclusions and have argued that HI can often be counterproductive and strategically selective (Weiss and Collins 1996). While the *droit d'ingérence* has been secured by the United States and her allies through the Security Council, Western-backed HI does not appear to have been balanced by a commitment to long-term conflict resolution and redevelopment.

In the longer term, the role and scope of HI will have to be clarified at the very least in terms of the deployment and development of the humanitarian roles of the armed forces, the influence of television coverage and the role of non-state actors (P. Taylor 1997). For sceptics, the militarization of HI provokes further levels of violence in crisis-ridden regions and complicates the process of reconciliation and rebuilding, whereas proponents argue that it may be necessary in the face of rampant human rights abuses and genocidal violence. In the absence of an effective government, as in Somalia, the provision of militarized HI could be justified in terms of seeking to reduce suffering in the face of armed militias imperilling citizens. As UN Secretary General Kofi Annan noted in 1993, 'The reality is there are situations when you cannot assist people unless you are prepared to take certain military measures' (cited in Weiss 1994: 6).

The role of non-state organizations such as the Red Cross also raises issues concerning the management of HI. Transnational networks of humanitarianism are changing the remits of provision and organization, and they make the promotion of non-statist and non-military forms of HI a pressing challenge for both Western and non-Western critics. The ability of NGOs and social movements to create a new moral consensus on humanitarian assistance should not be overrated, as the globalization of humanitarianism is still a long way off, judging by the lack of intervention in places such as Rwanda, Angola and Liberia as well as the opposition to HI from countries such as China (Pugh 1996; Prunier 1995; Shaw 1996).

Conclusion

Since the Second World War there has been a considerable extension of international law regarding moral standards of governance. The Universal Declaration of Human Rights remains highly significant in terms of establishing the specific responsibilities of governments *vis-à-vis* their peoples. More recently, the possibility of universal human rights has been advanced by some critical commentators to suggest that a global civil society could emerge to establish new sets of relationships between the individual, the state and the world. In the post-Cold War era, the norms of non-intervention and state sovereignty have been challenged by events in Europe and beyond. Old debates about national interests and self-interested intervention are required to coexist with new agendas based on collective security, common humanity and human rights protection.

Debates over humanitarian intervention and human rights will continue into the next century. Any criteria for judging the legitimacy of intervention will have to consider the nature of the authority approving the intervention; the motivation for the intervention and the outcome of the intervention. Difficult issues will have to be confronted, such as how one judges the nature and extent of the suffering which might justify intervention in cases where there is no proven record of massive human rights violations (Bretherton 1996). Television coverage of human suffering does not necessarily improve the capacity of decision makers to adjudicate on the nature of particular violations. Besides that, television coverage of human rights violations can be partial or it can ignore events entirely, as perhaps in East Timor, Angola, Mozambique and Kashmir (Chapter 3). In future the role of the military in the deliverance of HI needs to be carefully assessed, because the experience of the Somalian operation demonstrated that military-based intervention can have a disastrous impact on local and often vulnerable societies.

Human rights and HI are therefore intimately connected to the political and economic globalization of the planet. The provision of humanitarian aid illustrates a growing trend in global politics towards intervention in the affairs of other states. The longer-term challenge for proponents of HI is not only to establish a clear criterion for this endeavour but also to recognize that crises in places such as Bosnia and Somalia are products of a highly unequal global system which contributes to local and regional political conditions. Regions such as Central America and sub-Saharan Africa have witnessed worsening poverty, and Nicaragua has experienced a considerable drop in per capita income since the 1960s. Global solutions to humanitarianism will depend on adequately addressing issues relating to warfare, poverty, maldevelopment and despair.

Further reading

On human rights see *Understanding Human Rights* edited by C. Gearty and A. Tomkins (Mansell, 1996). On humanitarian intervention see *Humanitarian Intervention* edited by O. Ramsbotham and T. Woodhouse (Polity, 1996); A. Roberts' article 'Humanitarian war: military intervention and human rights', *International Affairs* **69**: 429–49, T. Weiss's 'UN responses in the former Yugoslavia: moral and operational choices', *Ethics and International Affairs* **8**: 1–22; and T. Weiss and L. Collins' *Humanitarian Intervention in the Post-Cold War Era* (Lynne Rienner, 1996). A classic if older study of intervention in international society is R.J. Vincent's *Non-Intervention and International Order* (Routledge & Kegan Paul, 1974).

Chapter 8

Conclusion

The challenge facing students of geopolitics is the ability to make geographical and political connections between an often bewildering series of stories and events. How, for example, might a news story about starving children in Sudan and Mali be connected to the offer of the International Monetary Fund (IMF) to provide financial assistance to Thailand and Russia in 1998? The answer could be that the international community led by the United States has tended to prioritize the development needs of strategically significant nations such as Russia at the expense of sub-Saharan Africa. It might also be noted that the IMF in the 1980s and 1990s encouraged Sudan to grow cotton as a major exporting crop (in order to raise money for debt repayment) rather than food crops such as maize (Cook 1994). When the world's cotton markets collapsed due to economic recession, Sudan was left indebted and there was little fertile land left for growing food, hence people starved. The ongoing and bitter civil war (started during the Cold War) further devastated many parts of Sudan (Halliday 1989; Hanlon 1996).

Underlying these sorts of media story is a sense in which events in sub-Saharan Africa, the United Kingdom and Russia are related to a range of transnational networks, processes and relationships. Sometimes these connections can be quite subtle; consider this story reported by the Brazilian media about the businessman Geritz Aronson. He was kidnapped in São Paulo by a group of well-known Brazilian criminals who demanded a US$1 million ransom for his safe release. However, these criminals were keeping abreast of current events, and on learning that the financial crisis had adversely affected Senor Aronson's electrical businesses, they reduced their demand to US$100,000 in view of the parlous state of the Brazilian economy. The money was duly deposited by a representative of the Aronson family, and the unfortunate captive was released in October 1998. After his release, a visibly relieved Senor Aronson declared on prime-time television that it was highly commendable that these robbers showed so much concern for the state of the national economy and his electrical companies! This tale demonstrates how a local kidnapping in São Paulo was ultimately influenced by the interaction of a series of local companies, the national economy and the ongoing machinations of global financial and information networks.

Many academic geographers and other scholars argue that economic and political forms of globalization are profoundly changing the nature of modern political and social life (Corbridge, Martin and Thrift 1994; Frost 1996; Falk 1997; Lauterpacht 1997; Ó Tuathail 1997; Linklater 1998). For students of geopolitics, the significance of an increasingly interconnected world is several-fold:

- Events in one part of the world may have profound implications for other parts of the world, regardless of distance and territory.
- The control of territorial space has become more problematic as flows of people, money and ideas challenge the ability of states to control these movements. Ideas of national sovereignty and territorial jurisdiction are becoming increasingly problematic.
- Globalization is producing a more unequal and hierarchial world in which some states are better equipped than others to take advantage of trade opportunities and/or resist economic recession.
- The distinction between the 'global' and the 'local' becomes harder to sustain in the light of transnational flows and networks.

One example of this blurring between the 'local' and the 'global' was provided in the summer of 1998 when Ealing council in west London announced it had no further housing available to accommodate fleeing Kosovan refugees. As a council located close to London's Heathrow airport, Ealing Housing Department has had to find emergency food and accommodation for refugees from the former Yugoslavia and other crisis-ridden regions such as Central Africa. Under existing British law and the 1951 Geneva Convention concerning refugees, local councils have an obligation to house and shelter refugees until a decision is made whether or not they can remain in the United Kingdom. Since this issue came to light, local newspapers and television stations have devoted more attention to the plight of Kosovo, and this has revealed how seemingly 'distant' events can affect the 'local' Ealing residents. At times this can be negative, with some residents complaining that local taxation will have to be raised in order to pay for the emergency provision of food and housing. Other residents acknowledge with some pride that the people of Ealing are contributing to the humanitarian endeavours of the international community by providing shelter for displaced peoples. These local expressions of civic pride are a direct result of events in the former Yugoslavia.

In this book I have tried to demontrate how issues such as nuclear proliferation, humanitarianism and financial and informational flows can sustain particular representations of a 'borderless world'. The dangers posed by a nuclear war, for example, clearly have implications for all living beings, regardless of their location. Moreover, the geographies of nuclear testing were such that particular communities have suffered and continue to suffer despite the absence of a global nuclear conflict. Appeals to global humanitarianism have placed new pressures on the nation state, forcing governments to heed the safety and well-being of their own citizens and vistors. However, it is also apparent that the United Nations has been prepared to react more strongly to some human rights abuses, e.g. Kosovo, than to others, e.g. Angola, Mozambique and areas in the Russian Federation (Harriss 1995; Hoogvelt 1997).

Figure 8.1 Anti-racist graffiti in Bruges, Belgium. In the last twenty years many Western European countries such as Belgium, France, Germany and the United Kingdom have experienced rising prejudice against immigrants and asylum seekers. The government and protest groups have sought to resist xenophobic parties such as the National Front in the United Kingdom. (*Photo*: Klaus Dodds)

I have also explored how this apparently 'borderless world' is a very unequal world, where national governments (such as the United States and Europe) attempt to restrict the movements of people (from Latin America and North Africa) through immigration policies (Figure 8.1) and border controls (Collinson 1996; Doty 1996; Mazower 1998). It is also a highly unequal environment in the sense that some states, e.g. Japan, China and the United States, enjoy far greater influence over global events and processes (including new trade regulations concerning investment and commerce) than entire regions such as sub-Saharan Africa and Central America. Unrestricted movement of people and capital remains a comparative luxury for those who are located in the South. The challenge for students of geopolitics is to chart these developments and then attempt to imagine alternative views of global change which include some sense of justice for those less fortunate than the wealthy minorities in both the North and the South (Simai 1997).

An unequal and interconnected world: final reflections

Almost a hundred years earlier, the British political geographer Halford Mackinder (1904) also argued that the ending of the nineteenth century would bring forth a different type of political and economic world (Schmidt 1998). International politics would henceforth be operating in a closed world system because, among other things, European colonialism and imperialism would encompass the entire earth's surface (Figure 8.2). New technologies such as the aeroplane, the telegraph and the telephone contributed to time–space compression and transformed the world into a functional geopolitical whole. Later claims to a new world order in the 1990s need to be considered carefully because assertions of uniqueness may either be misplaced or historically naive (Thrift 1992, 1995). There is no doubt that Victorian observers were convinced that massive economic, political and cultural change also occurred in the 1890s. Nineteenth-century ideas of 'circulation' were used by European writers to propose that the railway created new spaces of continuous movement and modern life. Life was becoming faster and there were fears that a kind of general hysteria (time–space depression) would characterize society. The major difference between then and now may be one not only of technological and social intensity (communication technologies have speeded up information circulation) but also geographical coverage (Shelley 1993).

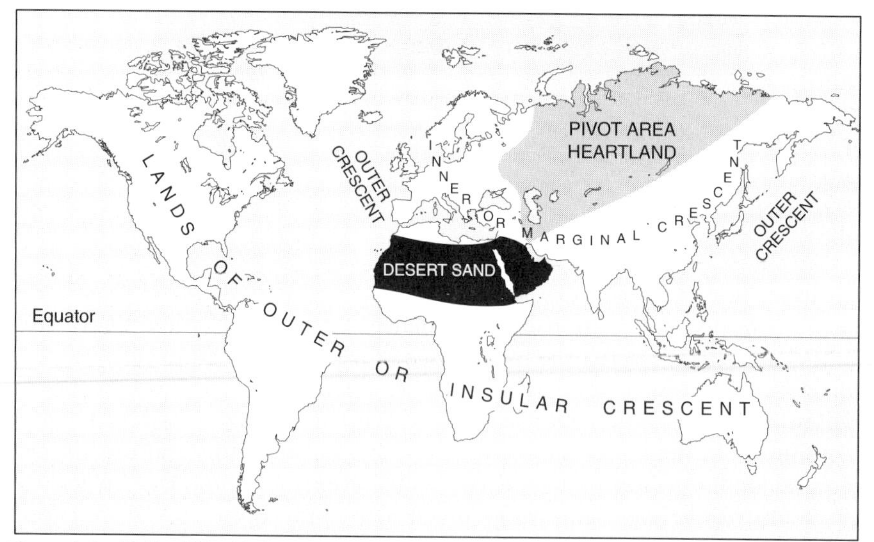

Figure 8.2 Halford Mackinder's 'heartland model' sought to depict the geopolitical and cultural significance of the Russian steppes for the future shape of human history. The map was first produced in 1904 and has been the most important image associated with formal geopolitics.

For some commentators, ideas such as frontier, boundary and movement became emblematic of a century characterized as an 'age of migration' (Castles and Miller 1993). As James Clifford once noted:

> This century has seen a drastic expansion of mobility, including tourism, migrant labour, immigration and urban sprawl. More and more people 'dwell' with the help of mass transit, automobiles, air planes [and trains]. In cities on six continents foreign populations have come to stay – mixing in but often in partial, specific fashions. The 'exotic' is uncannily close. Conversely, there seems no distant place left on the planet where the presence of 'modern' products, media, and power cannot be felt. An older topography and experience of travel is exploded. One no longer leaves home tackling something radically new, another time or space. Difference is encountered in the adjoining neighbourhood, the familiar turns up at the ends of the earth. (Clifford 1988: 13)

This book has been concerned throughout with globalization and the various ways in which it has promoted a greater interconnectivity between different places and peoples. It implies, therefore, that our daily lives (whether we like it or not) are being increasingly influenced by events and places beyond our local worlds (Allen and Massey 1995). Globalization is therefore both profoundly geographical and sociological in the sense that the stretching of social relationships occurs across the fixed boundaries of nation states. These relationships are also multidimensional (environmental, strategic, humanitarian, economic, political and cultural) and link various agents and structures over large distances (Figure 8.3). While there may be some disagreement over the origins of globalization, most commentators accept that experiences of globalization have become more intense for nation states, national economies and citizens in the last fifty years.

From a geopolitical viewpoint, contemporary discussions of globalization point to how the political geography of international relations is being challenged in three interlinking contexts. First, the growing interconnectedness of the world has imposed challenges on the theory and practice of state sovereignty; for an Australian example see Higgot (1997) and Higgot and Nossal (1997). The growth of supranational and transnational organizations is one indication of this transformation as states recognize that their sovereignty may be constrained by other actors. The activities of groups such as Greenpeace have played a part in challenging state authority by their exposure of sea-based dumping in the high seas, overfishing in the oceans and/or other forms of environmental degradation. Second, the movement of people across national boundaries is raising significant issues concerning sovereignty and citizenship. The flow of people across the US–Mexico border (see Figure 8.4) has demonstrated that state regulatory power is limited, in spite of American attempts to prevent unregulated movement. Recent controversies over the rights of so-called illegal aliens in California and Texas merely reflect this inability to control and regulate human movement absolutely. Another recent example was launched by the US Immigration Service in 1997, code-named Operation Gatekeeper, in an attempt to halt the flow of illegal immigrants from Mexico.

Figure 8.3 The Western Wall of Jerusalem. On the lower placard, the three notices are written in English, Hebrew and Arabic. The status of this disputed city has been a cause of concern for Arab, Christian and Jewish communities. The upper sign for the Misgav Ladach Road has had the Arabic notice defaced. (*Photo*: Klaus Dodds)

Using dog patrols, surveillance technologies and four-wheel drive vehicles, the service struggles to police a border that stretches 2,000 miles along the Rio Grande. In 1998 it was estimated that around 170 people had died trying to cross the border through heat exhaustion and drowning. The Binational Human Rights Centre in Tijuana believes that over 300,000 Mexicans per year illegally enter the United States in search of low-paid work such as fruit picking and domestic labour. One of the enduring ironies of this situation is that the free trade agreements signed between the United States, Canada and Mexico have freed the movement of capital in North America and simultaneously encouraged the United States to spend more money in order to prevent illegal immigration.

With the increased recognition that states are natural and/or absolute entities, hence existing territorial borders are only possible in terms of community

Figure 8.4 US-Mexico border crossing point.

and rights. The battle to control the border is thus central to sovereignty discourses, circumscribing the state's right to establish moral and political rights and responsiblities within the territorial boundaries. The condition of interconnectedness is imposing tension and uncertainty on national political decision-making systems. So-called transnational network societies have begun to emerge, and exist uneasily alongside the nation state and other organizations with a territorial basis. Social life is henceforth considered to be characterized by interconnectivity and hybridization across and through a range of geographical scales from the local to the global. Places and peoples are linked together in a relational myriad of multiple and asymmetric interdependencies, and it becomes impossible to consider the local, national, regional and global as separate spheres of social and territorial organization.

However, few of these issues are worldwide in scope, as different nations are affected to varying degrees. Terms such as 'global politics' can be misleading in the sense that they imply a certain uniformity rather than stressing the uneven and unequal nature of global processes and relations (Allen and Massey 1995). Moreover, in spite of claims to the contrary, nation states remain powerful actors in world politics and are not likely to be replaced by other forms of political organization in the first part of the next century. The experiences of globalization are not equal (and never have been) and this greater sense of interconnectivity may be negative rather than positive. For citizens in the old democracies of the North, cultural forms of globalization may be very positive; their supermarkets are now stocked with freshly picked strawberries from the United States, cheddar cheese from Australia and fish from the Southern Ocean. Globalization in this sense often implies more consumer choice. Alternatively, processes associated with time–space compression have had more disturbing consequences. The nuclear explosion at Chernobyl, Ukraine, in April 1986 led to consumer boycotts of meat and milk products because of fears that nuclear radiation had contaminated livestock in Britain. Greater interconnectivity can be a double-edged sword as conceptions of 'near and far' and 'risk' are overturned in the late twentieth century (Beck 1992).

In the South, personal and collective experiences of globalization can also be both positive and negative. It has been argued that Southern economies have greater opportunities to integrate themselves into a global economy based on free trade and market choice. On the other hand, globalization has often meant that countries such as Vietnam or Yemen are now more vulnerable to the unequal flow of goods, capital and ideas between the rich North and the poor South. The experiences of financial globalization, informational globalization and cultural globalization are always unequal and incomplete. As Allen and Massey have argued: 'there is little point in pretending that we are talking about the whole world here. Each "globalization" is constructing a different world' (1995: 3).

The principal contribution of geographical perspectives is not only to highlight the geographical uneveness but also the consequences for particular places. Four final points remain:

- The diffusion of communication technologies such as the telephone and the Internet have largely bypassed places such as sub-Saharan Africa and Central Asia in terms of direct access to the communication networks.
- Globalization has not meant that territorial space has been homogenized or transcended. Cultural diversity remains a welcome feature of global politics, and global products such as films, computer technologies and clothing are consumed differently. Ironically, globalization may well have contributed to a greater sense of difference between regions, ethnic groups and religious affiliations.
- Globalization has not meant that place, distance and geographical space have become irrelevant in world politics. The arguments over chronopolitics replacing geopolitics are frequently based on exaggeration instead of admitting there exists an interrelationship between territorial and supraterritorial geographies. For many people around the world, place remains significant as a source of national identity and natural resources, apparently confirmed by the recent wars over Bosnia and Kosovo.
- Globalization and the evolution of a 'borderless world' are not a panacea for dealing with chronic inequalities. Access to this borderless world is highly skewed in favour of the North in terms of trading opportunities, information generation, access to credit and control of global institutions such as the United Nations.

New states, new relations, new institutions and new networks

A world of greater interconnection and internationalization does not preclude local and regional diversity. This book has sought to explore some of these political, social and economic revolutionary changes (in the twentieth century). Throughout this account, attention has been paid to geographical differences within and between continents, regions, localities and the nation state. We continue to live in a world of enormous inequalities and subject to a bewildering range of changes. In the last ten years we have witnessed the unprecedented creation of new states in the former Soviet Union, Yugoslavia and parts of Africa. In August 1998 it was announced that St Kitts in the Caribbean had become the newest nation state after a plebiscite voted for cessation from its neighbour St Nevis. Meanwhile, the peoples of Diego Garcia, Eritrea and East Timor continue to struggle for independence, justice and self-determination. The former British colony Diego Garcia witnessed the displacement of the indigenous Ilois people from the islands in the 1960s so that the Americans could establish a military base in the Indian Ocean. For the last thirty years these displaced peoples have been told they cannot return from their temporary homes in Mauritius to Diego Garcia. Moreover, the island is now a nuclear weapons dump, and the American armed forces seek to extend their lease on the island following its use during Operation Desert Storm in 1991.

New relationships between states and other organizations have also arisen in the face of global issues such as environmental affairs, migration, AIDS disease and humanitarian emergencies. New institutions such as the World Trade Organization (WTO) combined with the growing visibility of pressure groups and transnational corporations have emerged to further complicate the networks of global governance. Questions of accountability and effectiveness are leading to renewed interest in the creation of new forms of rule-based governance for the twenty-first century (Herod, Ó Tuathail and Roberts 1998). Relatively new issues such as transboundary crime, drugs pushing and money laundering are also assuming a greater significance on the global political agenda.

In the 1980s and 1990s many governments, including the United States and the United Kingdom, attached greater importance to controlling the illict trade in drugs like cocaine and heroine and the subsequent profits accrued by drug cartels in places such as Colombia and Bolivia. In Latin America, the United States provided financial and military support to various governments as part of an ongoing project dedicated to destroying the drug trade. The United States remains one of the largest markets, not least because of the demand for high-grade cocaine among relatively affluent Americans. Sadly, many highly indebted economies such as Bolivia are estimated to generate far greater sums of money and employment from illegal drug exports than from the existing activities in tin mining (Green 1995). Proceeds from drug deals deposited in offshore bank accounts in places such as the Bahamas and the Cayman Islands are often beyond the reach of US or other national authorities (Stares 1996).

The international community's attempts to control drugs trading date back to an international conference on the opium trade in 1909. In the post-war period, the United Nations established various conventions, such as the

1961 Single Convention on Narcotic Drugs and the 1988 Convention Against the Illict Traffic in Narcotic Drugs and Psychotropic Substances, in order to curb the global movement of substances like cocaine, cannabis and opium. The United Nations has also established an International Drug Control Programme in Vienna, which seeks to coordinate international action on drug abuse control and the suppression of drug trafficking.

For students of global politics the implications are substantial, as all these trends and issues point to the changing subject matter of 'geopolitics'. The capacity of states to determine particular courses of action have been changed or even undermined by a range of transboundary and supraterritorial relationships. Electronic flows of money, information networks and environmental problems such as ozone depletion, all transgress territorial boundaries. The realist view of world politics based on states, national territory and formal models of sovereignty appears ill-equipped to understand the contemporary topographies of world politics. Moreover, the realist assumption that all states enjoy the same rights and responsibilities in the international arena look absurd in the face of evidence that some states depend on displays of international recognition for their continued legitimacy. A prime example here would be the new Kabila government of the Democratic Republic of Congo (formely Zaire), which is currently embroiled in a large-scale civil war with rival factions and neighbouring states such as Namibia, Uganda, Rwanda and Zimbabwe.

The major challenge facing political leaders, activists and a host of other interest groups is to imagine a world in which processes such as immigration, economic reform, human rights, humanitarian emergencies, environmental degradation and nuclear disarmament are handled in a manner which creates a safer and more just world. Whether our current system of nation states and intergovernmental organizations can produce such programmes and policies is a moot point. Perhaps we will have to think in new ways that explore different forms of political organization and cultural affiliation. The realist world of the sovereign state and national jurisdiction may well have had its moment in human history. In the meantime, students of geopolitics should retain a sense of humanity, justice and commitment for those oppressed, tortured and deprived of basic human or community rights. According to Michel Foucault, geographers who belong to the Euro-American world have a privilege that is denied to many students and scholars working in Nigeria, the former Yugoslavia, Mynamar (Burma), Iraq and East Timor – they can 'speak truth to the face of power'.

Further reading

On various themes mentioned in the conclusion, see T. Athanasiou's *Slow Reckoning: The Ecology of a Dying Planet* (Secker and Warburg, 1997), S. Castles and M. Miller's *Age of Migration* (Macmillan, 1993) and J. Pilger's *Secret Agendas* (Verso, 1996). On drugs see P. Stares' *Global Habit: The Drug Problem in a Borderless World* (Brookings Institution, 1996).

Popular geopolitics and beer

Equipped with your knowledge of critical geopolitics, try to deduce whether the beer is of Greek or Turkish origin, and say how you worked it out. The main clue lies in the map at the centre of the beer label. Turn over to check your answer.

The beer is produced by the Keo Beer Company based in Limassol, Cyprus. The map of Cyprus is geopolitically significant because it depicts an island undivided. Since the Turkish invasion of Cyprus in 1974, successive Greek governments have refused to accept the legitimacy of the UN-sponsored peace divide. Hence Greek companies, including Keo, produce maps of Cyprus that do not acknowledge the Turkish occupation of northern Cyprus.

Glossary

This short glossary is intended to offer some definitions concerning key terms such as geopolitics, globalization and new world order. These entries should be regarded as suggestive rather than definitive. For further details on key geographical terms see *The Dictionary of Human Geography*, 3rd edn, edited by R. Johnston, D. Gregory and D. Smith (Blackwell 1994) and *The Dictionary of Geopolitics*, edited by John O'Loughlin (Greenwood Press 1994).

Berlin Wall: The wall constructed by the Soviets to divide the city of Berlin in Germany into two sectors in 1961. After the Second World War, Berlin was occupied by four powers and their administrative sectors: France, the United Kingdom, the United States and the Soviet Union. After a series of crises, the Soviets decided to build a concrete wall across the centre of the city. It became a powerful illustration of a divided city and a European continent split between the capitalist West and the communist East. In November 1989, Germans toppled the East German regime and dismantled parts of the wall. Germany was reunified in 1990 and the capital has been moved from Bonn to Berlin.

Cold War: A term invented by the US journalist Walter Lippman to refer to the conflict and tension 1945–1991 between the United States and the Soviet Union (called the superpowers because of their military strength). The term 'Cold War' was popular because it implied a 'frosty' relationship between the two countries rather than outright war. However, the implications for the countries in the Third World were dramatic as superpower rivalry either worsened civil wars or provoked new conflicts. Nor did Europe escape from the violence of the Cold War, as rebellions against Soviet rule in Hungary (1956) and Czechoslovakia (1968) were ruthlessly crushed. The ending of the Cold War is usually dated from the fall of the Berlin Wall in November 1989 and the subsequent dismantlement of the Soviet Union in 1991.

Critical geopolitics: This term refers to a recent body of literature in North America and Europe which explored the geographical assumptions and understandings underpinning foreign policy-making and theories of world politics. Particular attention has been given to the use of geographical metaphors (e.g. heartland, containment, domino effect during the Cold War) and their significance in popular and formal geopolitics. Critical geopolitics has demonstrated that geopolitical themes are to be found in the cinema, newspapers, television and in music. In contrast to **geopolitics**, explanations are sought to determine how geographical labels and designations enter into popular and formal discourse rather than to imply a strong causal relationship between global physical geography and state behaviour.

Gender: This refers to the assumptions placed upon and the divisions made between men and women. It is not the same as 'sex', which soley refers to the biological differences between men and women. Recent work in gender and world politics has revealed that economic and political restructuring of the world economy has very different implications for men and women, not least because women are often expected to care for children while trying to work.

Geopolitics: Originally coined at the turn of the twentieth century, it referred to a particular approach to world politics which stressed the significance of territory and resources. During the post-war period, Anglo-American geographers were reluctant to use the term 'geopolitics' because they felt it had inspired Nazi Germany's policies of spatial expansionism. In the 1970s, however, political figures such as Henry Kissinger and Ronald Reagan used geopolitical language to describe international affairs and the Cold War against the Soviet Union.

Globalization: This term is widely used in the social sciences to point to the intensification and geographical spread of international interaction. Ideas of globalization that have emerged from an academic context include interdependence, internationalization, interpenetration, modernization, time–space compression, universalism and integration. From a geographical perspective, the literature on globalization raises profound challenges, not least because some authors have argued that territorial space has become less important in shaping world affairs. The new millennium, according to some writers, will be characterized by the domination of global capitalism, itself sustained by endless images of a borderless world, virtual financial flows and the complete domination of physical space. However, these varied representations of globalization rest on a series of assumptions about the social and political world, assumptions that have yet to be verified.

Hegemony: The capacity of a particular political or cultural group to exercise control and perpetuate inequality through the deployment of particular ideas and practices instead of by force. Employing the ideas of the Italian thinker Antonio Gramsci, who was imprisoned between 1928 and 1935, many scholars have explored how citizens might actually support ideas and practices that either curtail their liberties or impose restrictions on others. Hegemony therefore implies more than just the dominant ideology of elites and **popular geopolitics** seeks to explore how ideas about global geopolitical space are imbedded in everyday life via the education system and media outlets.

Intergovernmental organization: An intergovernmental organization (IGO) is usually a group of states which have created a governing body to manage an aspect of international affairs. A military example is the North Atlantic Treaty Organization (NATO) created in 1949 by sixteen states to coordinate the defence of the North Atlantic against Soviet forces during the Cold War. Far from dissolving with the Cold War, NATO has assumed considerable influence in the post-Cold War world and now plays a significant role in supporting UN operations, as witnessed in Bosnia. At the end of the twentieth century, NATO is attempting to expand its membership (under the Partnership for Peace Programme) to include Eastern European countries such as Poland, the Czech Republic and Hungary. Another intergovernmental organization is the International Labour Organization (ILO) of the United Nations. Created in 1919, it was formally adopted by the United Nations in December 1946. The ILO seeks to improve working conditions and to protect the rights of workers around the world in terms of health, safety and trade union membership. In 1969 the ILO was awarded the Nobel peace prize in recognititon of its contribution.

International Monetary Fund: Established by the Bretton Woods agreement in 1944 and operating since 1947 under the United Nations, the International Monetary Fund (IMF) advises governments and the **World Bank** on fiscal policies such as taxation, interest rate policy and the funding of public policy programmes.

New world order: Popularized, but not invented, by US President George Bush, the phrase was used in the early 1990s to describe the world at the end of the Cold War. It was hoped that this would be a moment for unprecedented international cooperation and a new opportunity for the United Nations to help govern a more peaceful world. Unfortunately, humanitarian crises in Iraq, Somalia, Bosnia and other parts of the world led some commentators to talk about a 'new world disorder' in the midst of such loss of life and homes.

Non-Aligned Movement: Created in 1961, this Southern political grouping is committed to five major principles: peace and disarmament, economic justice, self-determination, cultural respect and multilateral cooperation within bodies such as the United Nations. The political glue which bound the movement together was derived from a common desire to negotiate its own governance within the context of a superpower struggle and decolonization.

Non-governmental organization: A non-governmental organization (NGO) is a political organization which operates independently of states and other state-centric organizations. It often seeks to pursue radical political agendas independently from the formal realm of politics due to a belief that its goals should not be co-opted by mainstream politics. In terms of organizational structure, many NGOs are composed of flexible networks rather than rigid hierarchies.

North and South: These terms became increasingly popular with political observers in the 1980s. Many people in the Third World argued that, with the ending of the Cold War, the most fundamental differences were due to inequalities of wealth. The terms 'North' and 'South' are geographical in the sense that the wealthiest countries of the world tend to be located in the northern hemisphere and the poorest countries in the southern hemisphere.

Popular geopolitics: A term used in **critical geopolitics** to refer to the geographical representations of global political space found in popular cultural forms such as cartoons, films, novels and music. It is argued that popular culture plays a significant role in reproducing certain hegemonic values about political space, values that need to be carefully examined.

Structural adjustment programme: One of several programmes designed in the 1980s and 1990s by the IMF and the World Bank in an attempt to pressurize indebted Southern states such as Argentina, Zimbabwe and Thailand to undertake public sector savings in return for further loans and grants. In effect, it often forced the recipient government to cut its spending on health care, education and other public sectors, in order to achieve these targets.

Third World: A term invented by French social scientists in the 1950s to describe the continents of Africa, Asia, Latin America and Oceania. These parts of the world were also known as the 'developing world' because they were considered to be distinct from the advanced economies of the First World (sometimes called the North or the West) and the Second World of socialist states (sometimes called the East or the Communist bloc). Many writers prefer the term 'South' because it is thought that 'Third World' is derogatory to peoples living in the southern hemisphere.

Transboundary: The term 'transboundary' describes processes and phenomena which have an intrinsic capacity to cross territorial and other administrative boundaries. Two examples are money and acid rain.

Transnational corporation: A large organization that operates in a number of different national economies, it is by definition **transboundary** because it has a network of locations and activities, often a complex network. It is estimated that around 30 per cent of the world's trade is conducted between transnational corporations (TNCs), and this growing influence has been assisted by the development of communication and financial networks which allow money and trade to flow around the globe at an ever faster rate.

UNESCO: The United Nations Education, Scientific and Cultural Organization was created in November 1946 for the specific purpose of promoting collaboration among nations through educational, scientific, media and cultural projects. It has been at the forefront of promoting the free flow of information, improving educational opportunities in the South, maintaining a United Nations University in Tokyo, safeguarding places and sites of great ecological, cultural and historical importance, and raising the literacy rates of women. Women's literacy has been considered a crucial dimension in the various UN-sponsored programmes for improving the lives and conditions of women in the North and South. UNESCO has sometimes been a controversial body because it has called for fundamental reform in the methods of collecting information, exchanging it and disseminating it free of censorship and state interference.

World Bank: Created by the Bretton Woods agreement in 1944, the World Bank (formally the International Bank for Reconstruction and Development) was designed to provide loans and grants for economic development. Since its inception, it has lent US$333 billion (adjusted 1996 figures) for developmental projects. Famous World Bank projects include the massive dam construction programmes in Egypt (the Nasser Dam) and in Zambia (the Kariba Dam). Money for the World Bank is provided by rich industrialized countries such as the United States and wealthy oil-exporting states such as Saudi Arabia.

World systems theory: An approach to global change which stresses the importance of changing modes of production. Since the fifteenth century, a capitalist world economy has provided a powerful catalyst for social, economic and political change. The spatial organization of the world economy is divided into the core (major economies such as the United States, Japan and the United Kingdom), the semi-periphery (nations such as Argentina, Australia, Indonesia, New Zealand and South Africa) and the periphery (the nations of the South). These categories are not considered static but are the result of a particular set of geographical and historical outcomes. World systems analysts hope that by studying the 'big picture' they will be better able to predict whether it will be possible to change the unequal nature of the capitalist world economy. Most hope that a socialist (or at least more egalitarian) form of production might be possible. Other scholars are uncomfortable with the sweeping historical and geographical analyses of world systems analysis.

World Trade Organization: The World Trade Organization (WTO) was created in 1994 following a lengthy series of trade negotiations known as the Uruguay Round. Its job is to ensure the principles of free trade and fair competition are upheld in the world economy. Critics in the South complain that the WTO will not have sufficient power to prevent the major trading economies, such as the United States and China, from pushing an agenda that exposes fragile economies to unregulated competition along with minimum controls on worker rights and environmental protection.

Bibliography

Afshar, H. (ed.), 1998, *Women and Empowerment*, Basingstoke: Macmillan.
Agnew, J., 1992, The US position in the world geopolitical order after the Cold War, *Professional Geographer* **44**: 7–10.
Agnew, J., 1994, The territorial trap: the geographical assumptions of international relations theory, *Review of International Political Economy* **1**: 53–80.
Agnew, J., 1998, *Geopolitics*, London: Routledge.
Agnew, J. and S. Corbridge, 1995, *Mastering Space*, London: Routledge.
Allen, J. and D. Massey (eds), 1995, *Geographical Worlds*, Oxford: Oxford University Press.
Arnold, G., 1993, *The End of the Third World*, Basingstoke: Macmillan.
Athanasiou, T., 1997, *The Ecology of a Dying Planet*, London: Secker and Warburg.
Ayoob, M., 1993, The new–old disorder in the Third World, *Global Governance* **1**: 59–78.
Ayoob, M., 1995, *The Third World Security Predicament: State Making, Regional Conflict, and the International System*, Boulder CO: Lynne Rienner.
Barker, C., 1998, *Global Television*, London: Routledge.
Barton, J., 1997, *A Political Geography of Latin America*, London: Routledge.
Baylis, J. and S. Smith (eds), 1997, *The Globalization of World Politics*, Oxford: Oxford University Press.
Beck, U., 1992, *The Risk Society*, London: Sage.
Berger, M., 1994, The end of the Third World, *Third World Quarterly* **15**: 257–75.
Block, F., 1996, *The Vampire State and Other Stories*, New York: New Press.
Bretherton, C., 1996, Introduction: Global politics in the 1990s, in C. Bretherton and G. Ponton (eds) *Global Politics*, Oxford: Blackwell, pp. 1–20.
Brown, C., 1997, *Understanding International Relations*, Basingstoke: Macmillan.
Bull, H., 1965, *The Control of the Arms Race*, London: Methuen.
Bull, H., 1977, *The Anarchical Society*, London: Macmillan.
Boutros-Ghali, B., 1992, *An Agenda for Peace*, New York: United Nations Publication Unit.
Bowman, I., 1921, *The New World*, New York: World Books.
Burchill, S. and A. Linklater (eds), 1996, *Theories of International Relations*, Basingstoke: Macmillan.
Bush, E. and L. Harvey, 1997, Joint implementation and the ultimate objective of the UN Framework Convention on Climate Change, *Global Environmental Change* **7**: 265–86.
Castaneda, J., 1994, *Utopia Unarmed*, New York: Vintage.
Castells, M., 1996, *The Rise of the Network Society*, Oxford: Blackwell.
Castles, S. and M. Miller, 1993, *The Age of Migration*, Basingstoke: Macmillan.

Chaturvedi, S., 1996, *Polar Regions: A Political Geography*, Chichester: John Wiley.
Chaturvedi, S., 1998, Common security? Geopolitics, development, South Asia and the Indian Ocean, *Third World Quarterly* **19**: 701–24.
Chomsky, N., 1991, *Deterring Democracy*, London: Verso.
CIIR, 1997, *Comment. East Timor: The Continuing Betrayal*, London: Catholic Institute for International Affairs.
Clark, I., 1997, *Globalization and Fragmentation*, Oxford: Oxford University Press.
Clifford, J., 1988, *The Predicament of Culture*, Berkeley CA: University of California Press.
Cohn, C., 1987, Sex and death in the rational world of defense intellectuals, *Signs* **12**: 687–718.
Collinson, S., 1996, *Shore to Shore: The Politics of Migration in Euro-Maghreb Relations*, London: Royal Institute of International Affairs.
Commission on Global Governance, 1995, *Our Global Neighbourhood*, Oxford: Oxford University Press.
Cook, I., 1994, New fruits and vanity: symbolic production in the global political food economy, in A. Bonanno, A. Busch, L. Friedland, W. Gouveia and E. Mingione (eds) *From Columbus to ConAgra*, Lawrence KA: University of Kansas.
Corbridge, S., R. Martin and N. Thrift (eds), 1994, *Money, Space and Power*, Oxford: Blackwell.
Cosgrove, D., 1994, Contested global visions: one world, whole earth and the Apollo space photographs, *Annnals of the Association of the American Geographers* **84**: 270–94.
Crang, M., 1998, *Cultural Geography*, London: Routledge.
Cresswell, T., 1996, *In Place/Out of Place*, Minneapolis: University of Minnesota Press.
Dalby, S., 1990, *Creating the Second Cold War*, London: Belhaven Press.
Dalby, S., 1991, Critical geopolitics: discourse, difference and dissent, *Environment and Planning D: Society and Space* **9**: 261–83.
Danchev, A. (ed.), 1995, *Fin de Siecle*, London: Macmillan.
Der Derian, J., 1992, *Anti-Diplomacy*, Oxford: Blackwell.
Desai, V. and R. Imrie, 1998, The new managerialism in local governance: North–South dimensions, *Third World Quarterly* **19**: 635–50.
Dickins, P., 1992, *Global Shift*, London: Paul Chapman.
Dines, G., 1995, Towards a sociological analysis of cartoons, *Humor* **8**: 237–55.
Dobson, A., 1990, *Green Political Thought*, London: Routledge.
Dodds, K., 1996, The 1982 Falklands War and a critical geopolitical eye: Steve Bell and the If . . . cartoons, *Political Geography* **15** (6/7): 571–92.
Dodds, K., 1997, *Geopolitics in Antarctica: Views from the Southern Oceanic Rim*, Chichester: John Wiley.
Dodds, K., 1998, Political geography I: the globalization of world politics, *Progress in Human Geography* **23**: 595–606.
Dodds, K., 1999, *An Ocean in Crisis? The Illegal, Unregulated Unreported Fishing of Patagonian Toothfish and the Southern Ocean*. CEDAR Discussion Paper Series, Royal Holloway, University of London.
Dorrance, J., 1985, ANZUS misperceptions, mythlogy and reality, *Australian Quarterly* (spring): 215–30.
Doty, R., 1996, *Imperial Encounters*, Minneapolis MN: University of Minnesota Press.
Doyle, T., 1998, Sustainable development and Agenda 21: the secular bible of global free markets and pluralist democracy, *Third World Quarterly* **19**: 771–86.

Doyle, T. and D. McEachern, 1998, *Environment and Politics*, London: Routledge.
Elliot, L., 1998, *The Global Politics of the Environment*, London: Routledge.
Enloe, C., 1989, *Bananas, Bases and Beaches: Making Feminist Sense of International Politics*, London: Pandora.
Enloe, C., 1993, *The Morning After: Sexual Politics at the End of the Cold War*, Berkeley CA: University of California Press.
Escobar, A., 1995, *Encountering Development*, Princeton NJ: Princeton University Press.
Falk, R., 1995, *On Humane Governance*, Oxford: Polity.
Falk, R., 1997, Will globalization win out? *International Affairs* **73**: 123–36.
Frank, A.G., 1971, *Capitalism and Under-Development in Latin America*, Harmondsworth: Penguin.
Freedman, L. and E. Karsh, 1993, *The Gulf War and the New World Order*, Basingstoke: Macmillan.
Frost, M., 1996, *Ethics in International Relations*, Cambridge: Cambridge University Press.
Fukuyama, F., 1992, *The End of History and the Last Man*, New York: Free Press.
Gamble, A. and Payne, A. (eds), 1996, *Regionalism and World Order*, Basingstoke: Macmillan.
Gardner, G., 1994, *Nuclear Nonproliferation*, Boulder CD: Lynne Rienner.
Gearty, C. and A. Tomkins (eds), 1996, *Understanding Human Rights*, London: Mansell.
Giddens, A., 1996, *Modernity and Self-Identity*, Cambridge: Polity.
Gilpin, R., 1987, *The Political Economy of International Relations*, Princeton NJ: Princeton University Press.
Gould, S., 1998, *Questioning the Millennium*, London: Vintage.
Gow, J., 1997, *Triumph of the Lack of Will: International Diplomacy and the Yugoslav War*, London: C. Hurst.
Graham, G., 1996, *Ethics and International Relations*, Oxford: Blackwell.
Grant, C., 1995, Equity in international relations: a third world perspective, *International Affairs* **71**: 567–88.
Green, D., 1995, *Silent Revolution*, London: Cassell.
Gupta, J., 1997, *The Climate Change Convention and Developing Countries*, Dordrecht: Kluwer.
Halliday, F., 1989, *Cold War, Third World*, London: Verso.
Hanlon, J., 1996, *Peace Without Profit: How the IMF Blocks Rebuilding in Mozambique*, Oxford: James Currey.
Harrison, P. and R. Palmer, 1986, *News Out of Africa*, London: Hilary Shipman.
Harriss, J. (ed.), 1995, *The Politics of Humanitarian Intervention*, London: Pinter.
Haynes, J., 1996, *Third World Politics*, Oxford: Blackwell.
Held, D., 1995, *Democracy and the Global Order*, Cambridge: Polity.
Hepple, L.W., 1986, The revival of geopolitics, *Political Geography Quarterly* **5**: 21–36.
Hepple, L.W., 1992, Metaphor, geopolitical discourse and the military in South America, in T. Barnes and J. Duncan (eds) *Writing Worlds*, London: Routledge, pp. 136–55.
Herod, A., G. Ó Tuathail and S. Roberts (eds), 1998, *An Unruly World*? London: Routledge.
Higgot, R., 1997, De facto and de jure regionalism: the double discourse of regionalism in the Asia-Pacific, *Global Society* **11**: 165–85.

Higgot, R. and K. Nossal, 1997, The international politics of liminality: relocating Australia in the Asia-Pacific, *Australian Journal of Political Science* **32**: 169–86.

Hirst, P. and G. Thompson, 1996, *Globalization in Question*, Cambridge: Polity.

Hobsbawm, E., 1997, *On History*, London: Weidenfeld and Nicolson.

Honig, J. and N. Booth, 1996, *Srebenica: Record of a War Crime*, Harmondsworth: Penguin.

Hoogvelt, A., 1997, *Globalization and the Post-Colonial World*, Basingstoke: Macmillan.

Huntington, S., 1993, The clash of civilisations, *Foreign Affairs* **72**: 22–49.

Huntington, S., 1996, *The Clash of Civilizations and the Remaking of World Order*, New York: Simon and Schuster.

Huque, M., 1997, Nuclear proliferation in South Asia, *International Studies* **34**: 1–14.

Hurrell, A., 1995, International political theory and the global environment, in K. Booth and S. Smith (eds) *International Relations Theory Today*, Cambridge: Polity, pp. 129–53.

Hurrell, A., 1998, Security in Latin America, *International Affairs* **74**: 529–46.

Imber, M. and J. Vogler (eds), 1995, *The Environment and International Relations*, London: Routledge.

Immerman, R., 1982, *The CIA in Guatemala*, Austin TX: University of Texas Press.

Jackson, R., 1990, *Quasi-States*, Cambridge: Cambridge University Press.

Johnston, R., D. Gregory and D. Smith (eds), *The Dictionary of Human Geography*, Oxford: Blackwell.

Kato, H., 1993, Nuclear globalism: traversing rockets, satellites and nuclear war via the strategic goze, *Alternatives* **18**: 339–60.

Kellner, D., 1992, *The Persian Gulf TV War*, Boulder CO: Lynne Rienner.

Kellner, D., 1995, *Media, Culture and Society*, London: Routledge.

Keohane, R. and J. Nye, 1989, *Power and Interdependence*, Boston MA: Harvard University Press.

Kolko, G., 1997, *Vietnam: Anatomy of Peace*, London: Routledge.

Kristof, L., 1960, The origins and evolution of geopolitics, *Journal of Conflict Resolution* **4**: 15–51.

Lacoste, Y., 1973, An illustration of geographical warfare: bombing of the dykes on the Red River, North Vietnam, *Antipode* **5**: 1–13.

Lacoste, Y., 1976, *La géographie, ça est, d'abord, à faire la guerre*, Paris: Maspero.

Lane, D., 1996, *The Rise and Fall of State Socialism*, Cambridge: Polity.

Latin American Outlook, 1995, Taking Mexico to market, speech delivered by Carlos Heredia to the Latin American Bureau, March 1995.

Lauterpacht, E., 1997, Sovereignty: myth or reality, *International Affairs* **73**: 137–50.

Leyshon, A., D. Matless and G. Revill, 1995, The place of music, *Transactions of the Institute of British Geographers* **20**: 423–33.

Lifton, R. and R. Falk, 1982, *Indefensible Weapons*, New York: Basic Books.

Lilley, R., 1997, Warships and planes hunt for Antarctic fish raiders, *Independent*, 1 May 1997.

Linklater, A., 1998, *The Transformation of Political Community*, Cambridge: Polity.

Lipschultz, R. and K. Conca (eds), 1993, *The State and Social Power in Global Environmental Politics*, New York: Columbia University Press.

Mackinder, H., 1904, The geographical pivot of history, *Geographical Journal* **23**: 421–44.

McCannon, J., 1998, *Red Arctic*, Oxford: Oxford University Press.

McGrew, A., 1998, The globalization debate: putting the advanced capitalist state in its place, *Global Society* **12**: 299–322.

Malcolm, N., 1998, *Kosovo: A Short History*, London: Macmillan.
Mazarr, M., 1995, *North Korea and the Bomb*, Basingstoke: Macmillan.
Mazower, M., 1998, *Dark Continent: Europe's Twentieth Century*, Harmondsworth: Penguin.
Mermin, J., 1997, US intervention and the new world order: lessons from the Cold War and the post-Cold War cases, *Political Studies Quarterly* **15**: 77–102.
Middleton, N., P. O'Keefe and S. Mayo, 1993, *Tears of the Crocodile*, London: Pluto.
Miller, M., 1995, *The Third World in Global Environmental Politics*, Milton Keynes: Open University Press.
Mittleman, J. (ed.), 1996, *Globalization: Critical Reflections*, Boulder CO: Lynne Rienner.
Morgenthau, H., 1948, *Politics Among Nations*, New York: Alfred Knopf.
Morley, D. and K. Robins, 1995, *Spaces of Identity*, London: Routledge.
Mullerson, R., 1996, *Human Rights Diplomacy*, London: Routledge.
Naisbitt, J., 1995, *Global Paradox*, New York: Avon Books.
Newhouse, J., 1989, *The Nuclear Age*, London: Michael Joseph.
Nijman, J., 1992, *The Geopolitics of Power and Conflict*, Chichester: John Wiley.
Nino, C., 1995, *Radical Evil on Trial*, New Haven CT: Yale University Press.
Norton-Taylor, R., 1998, Comic strip Animal Farm used as a Cold War weapon, *The Guardian* (17 March 1998).
Ohmae, K., 1990, *The Borderless World*, London: Collins.
O'Loughlin, J., 1989, World power competition and local conflicts in the Third World, in R. Johnston and P. Taylor (eds) *A World in Crisis*, Oxford: Blackwell, pp. 289–332.
O'Loughlin, J., 1992, Ten scenarios for a new world order, *Professional Geographer* **44**: 24–28.
O'Loughlin, J. (ed.), 1994, *The Dictionary of Geopolitics*, Washington DC: Greenwood Press.
Ó Tuathail, G., 1994, Displacing geopolitics: writing on the maps of global politics, *Environment and Planning D: Society and Space* **12**: 525–46.
Ó Tuathail, G., 1996, *Critical Geopolitics*, London: Routledge.
Ó Tuathail, G., 1997, Emerging markets and other simulations: Mexico, the Chiapas revolt and geopolitical panopticon, *Ecumene* **4**: 300–17.
Ó Tuathail, G., 1998, Postmodern geopolitics? the modern geopolitical imagination and beyond in G. Ó Tuathail and S. Dalby (eds) *Rethinking Geopolitics*, London: Routledge, pp. 16–38.
Ó Tuathail, G. and J. Agnew, 1992, Geopolitics and foreign policy: practical geopolitical reasoning in American foreign policy, *Political Geography* **11**: 190–204.
Ó Tuathail, G. and S. Dalby (eds), 1998, *Rethinking Geopolitics*, London: Routledge.
Ó Tuathail, G., S. Dalby and P. Routledge (eds), 1998, *The Geopolitics Reader*, London: Routledge.
Ó Tuathail, G. and T. Luke, 1994, Present at the (dis)integration: deterritorialisation and reterritorialisation in the new wor(l)d order, *Annals of the Association of American Geographers* **84**: 381–94.
Painter, J., 1995, *Politics, Geography and Political Geography*, London: Edward Arnold.
Palmer, J., 1992, Towards a sustainable future, in D. Cooper and J. Palmer (eds) *Environment in Question*, London: Routledge.
Parker, G., 1985, *Western Geopolitical Thought*, London: Croom Helm.
Parkins, C., 1996, North–South relations and globalization after the Cold War, in C. Bretherton and G. Ponton (eds) *Global Politics*, Oxford: Blackwell, pp. 49–73.
Paterson, M., 1996, *Global Warming and Global Politics*, London: Routledge.

Pettman, J., 1996, *Worlding Women*, London: Routledge.
Piel, G., 1992, *Only One World*, New York: United Nations.
Pilger, J., 1998, *Secret Agendas*, London: Vintage.
Pinochet, A., 1984, *Geopolitica*, Santiago de Chile: Andres Bello.
Pion-Berlin, D., 1989, *Ideology of State Terror*, Boulder CO: Lynne Rienner.
Princern, T. and M. Finger, 1994, *Environmental NGOs in World Politics*, London: Routledge.
Prins, G. (ed.), 1993, *Threats Without Enemies*, London: Earthscan.
Prunier, G., 1995, *Rwanda: History of a Genocide 1959–1994*, London: Hurst.
Pugh, M., 1996, Humanitarianism and peace-keeping, *Global Society* **10**: 205–24.
Ramsbotham, O. and T. Woodhouse, 1996, *Humanitarian Intervention in Contemporary Conflict*, Cambridge: Polity Press.
Redclift, M., 1987, *Sustainable Development*, London: Methuen.
Reyntiens, F., 1995, *L'Afrique des Grands Lacs, Rwanda, Burundi 1988–1994*, Paris: Les Afriques Karthala.
Rist, G., 1997, *History of Development*, London: Zed.
Roberts, A., 1993, Humanitarian war: military intervention and human rights, *International Affairs* **69**: 429–49.
Robertson, R., 1992, *Globalization: Social Theory and Global Culture*, London: Sage.
Robins, K., 1995, *Spaces of Identity*, London: Routledge.
Routledge, P., 1996, Critical geopolitics and terrains of resistance, *Political Geography* **15**: 509–32.
Routledge, P., 1998, Going globile: spatiality, embodiment and mediation in the Zapatista insurgency, in G. Ó Tuathail and S. Dalby (eds) *Rethinking Geopolitics*, London: Routledge, pp. 240–60.
Said, E., 1978, *Orientalism*, Harmondsworth: Penguin.
Said, E., 1993, *Culture and Imperialism*, London: Chatto and Windus.
Schmidt, B., 1998, *The Political Discourse of Anarchy: A Disciplinary History of International Relations*, Albany NY: State University Press of New York.
Scholte, J., 1997, *Globalization: A Critical Introduction*, Basingstoke: Macmillan.
Seager, J., 1993, *Earth Follies*, London: Earthscan.
Shafer, S. and A. Murphy, 1998, The territorial strategies of IGOs: implications for environment and development, *Global Government* **4**: 257–76.
Sharp, J., 1993, Publishing American identity: popular geopolitics, myth and the *Readers Digest*, *Political Geography* **12**: 491–503.
Sharp, J., 1996, Hegemony, popular culture and geopolitics: the *Readers Digest* and the construction of danger, *Political Geography* **15**: 557–70.
Shaw, M., 1996, *Civil Society and the Media in Global Crises*, London: Pinter.
Shelley, F., 1993, Political geography, the new world order and the city, *Urban Geography* **14**: 557–67.
Shiva, V., 1993, The greening of global reach, in J. Childs and J. Cutler (eds) *Global Visions: Beyond the New World Order*, Boston MA: South End Press.
Shohat, E. and R. Stam, 1994, *Unthinking Ethnocentrism*, London: Routledge.
Sidaway, J. and D. Simon, 1993, Geopolitical transition and state formation: the changing political geographies of Angola, Mozambique and Namibia, *Journal of Southern Africa Studies* **19**: 6–28.
Simai, M., 1997, The changing state system and the future of global governance, *Global Society* **11**: 141–64.
Simon, D. and Dodds, K. (eds), 1998, Rethinking geographies of North–South development, *Third World Quarterly* **19**: 593–791.

Simon, D., W. van Spengen, C. Dixon and A. Narman (eds), 1995, *Structurally Adjusted Africa*, London: Pluto.
Singham, A. and S. Hune, 1986, *Non-Alignment in the Age of Alignment*, London: Zed.
Slater, R., B. Schultz and S. Dorr (eds), 1993, *Global Transformation and the Third World*, Boulder CO: Lynne Rienner.
Smith, A., 1995, *Nations and Nationalism in a Global Era*, Cambridge: Cambridge University Press.
Smith, P., 1997, *Millennial Dreams*, London: Verso.
Smith, S., 1994, Soundscape, *Area*, **26**: 232–40.
Spybey, T., 1996, *Globalization and World Society*, Oxford: Polity.
Stares, P., 1996, *Global Habit: The Drug Problem in a Borderless World*, Washington DC: Brookings Institution.
Stoett, P., 1994, The environment enlightenment: security analysis meets ecology, *Coexistence* **31**: 127–47.
Stokke, O. and D. Vidas (eds), 1996, *Governing the Antarctic*, Cambridge: Cambridge University Press.
Taylor, P., 1997, *Global Communications, International Affairs and the Media Since 1945*, London: Routledge.
Taylor, P.J., 1993, *Political Geography*, Harlow: Addison Wesley Longman.
Taylor, P.J., 1996, *The Way the Modern World Works*, Chichester: John Wiley.
Thomas, C., 1987, *In Search of Security: The Third World in International Relations*, Brighton: Harvester.
Thomas, C. and P. Wilkins (eds), 1997, *Globalization in the South*, Basingstoke: Macmillan.
Thompson, E., 1994, *The Age of Extremes*, London: Verso.
Thrift, N., 1992, Muddling through: world order and globalization, *Professional Geographer* **44**: 3–7.
Thrift, N., 1995, A hyperactive world, in R. Johnston, P. Taylor and M. Watts (eds) *Geographies of Global Change*, Oxford: Blackwell: 18–35.
United Nations, 1996, *The United Nations and Somalia 1992–1996*, New York: United Nations Publications.
United Nations, 1998, *UN Human Development Report*, New York: Oxford University Press.
Vattimo, G., 1993, *The Transparent Society*, Cambridge: Verso.
Vincent, J., 1974, *Non-Intervention and International Order*, Princeton NJ: Princeton University Press.
Virilio, P., 1986, *Speed and Politics*, New York: Semiotext.
Virilio, P., 1989, *War and Cinema*, London: Verso.
Virilio, P., 1997, *Open Sky*, London: Verso.
Vogler, J., 1995, *Global Commons*, Chichester: John Wiley.
Vogler, J. and M. Imber (eds), 1995, *The Environment and International Relations*, London: Routledge.
Walker, R., 1988, *One World, Many Worlds: Struggles for a Just World Peace*, Boulder CO: Lynne Rienner and Zed.
Walker, R., 1993, *Inside/Outside*, Cambridge: Cambridge University Press.
Walker, W., 1998, International nuclear relations after the Indian and Pakistani test explosions, *International Affairs* **74**: 505–28.
Wallerstein, I., 1980, *The Modern World System II*, New York: Academic Press.
Waltz, K., 1979, *Theory of International Politics*, Reading MA: Addison-Wesley.

Ward, D., 1997, *Latin America*, London: Routledge.
Waters, M., 1995, *Globalization*, London: Routledge.
Weiss, T., 1994, UN responses in the former Yugoslavia: moral and operational choices, *Ethics and International Affairs* **8**: 1–22.
Weiss, T. and L. Collins, 1996, *Humanitarian Intervention in the Post-Cold War Era*, Boulder CO: Lynne Rienner.
Whitaker, J., 1997, *The United Nations*, London: Routledge.
White, B., R. Little and M. Smith (eds), 1997, *Issues in World Politics*, Basingstoke: Macmillan.
Whitfield, S., 1991, *The Culture of the Cold War*, Baltimore: John Hopkins University Press.
Willetts, P., 1978, *The Non-Aligned Movement*, London: Pinter.
Williams, M., 1997, The Group of 77 and global environmental politics, *Global Environmental Change* 7: 295–98.
Woodward, S., 1995, *Balkan Tragedy*, Washington DC: Brookings Institution.

Index